G000090757

Birds

of the Northwestern

National Parks

Number Forty-five,
The Corrie Herring Hooks Series

BIRDS
of the NORTHWESTERN
NATIONAL PARKS

A Birder's Perspective

Olympic National Park, Washington

North Cascades National Park, Washington

Mount Rainier National Park, Washington

Crater Lake National Park, Oregon

Oregon Caves National Monument, Oregon

Lava Beds National Monument, California

Lassen Volcanic National Park, California

By Roland H. Wauer

Drawings by Mimi Hoppe Wolf

University of Texas Press, Austin

Copyright © 2000 by Roland H. Wauer
All rights reserved
Printed in the United States of America
First edition, 2000

Requests for permission to reproduce material from this work
should be sent to Permissions, University of Texas Press, P. O.
Box 7819, Austin, TX 78713-7819.

The paper used in this book meets the minimum requirements
of ANSI/NISO Z39.48-1992 (R1997) (Permanence of Paper).

Library of Congress Cataloging-in-Publication Data
Wauer, Roland H.
    Birds of the northwestern national parks : a birder's per-
spective / by Roland H. Wauer ; drawings by Mimi Hoppe
Wolf. — 1st ed.
        p. cm. — (Corrie Herring Hooks series ; no. 45)
    Includes bibliographical references (p.) and index.
    ISBN 0-292-79133-x (pbk. : alk. paper)
    1. Birds—Pacific States. 2. Bird watching—Pacific States—
Guidebooks. 3. National parks and reserves—Pacific States—
Guidebooks. 4. Pacific States—Guidebooks. I. Title.
II. Series.
QL683.P37 W38   2000
598'.07'234795—dc21   99-055373

Contents

Preface

America's national parks possess the best examples of the continent's natural heritage, complete with the grandest scenery and most stable plant and animal communities still in existence. In a large sense, our national parks represent a microcosm of our last remaining wildlands.

Birds of the Northwestern National Parks: A Birder's Perspective has a three-fold purpose. First, it provides the reader with a fresh perspective on bird life in seven magnificent national park units within Washington, Oregon, and Northern California. It is intended to introduce the park visitor to the most common and obvious birds as well as to the fascinating world of bird identification and behavior. It can be used as a reference to the park and its bird life during a park visit, and as a valuable tool in preparing for that visit.

Second, *Birds of the Northwestern National Parks* provides detailed guidelines on how best to get started with the fascinating activity of bird-watching. It helps the reader select the best tools, offers keys to bird identification and field techniques, and also discusses birding ethics.

Third, and just as important, the chapter titled Parks as Islands contains a discussion on the status of our national parks. Sections titled Threats to the Parks, Wildland Fires and Their Effects on Bird Life, The Future, What Is Being Done within the Parks?, and The Value of National Parks contain this writer's perspective on the health and vitality of the national parks and the National Park System.

This book is not intended for use as a field guide or a book on bird identification per se. Several excellent field guides are already available; they should be used as companion volumes to this book. Nor is this book intended to help the birder find the rarities or the out-of-the-way specialties. Its purpose is to help the park visitor better appreciate the parks and their bird life. If, in making new acquaintances, the park visitor should become interested in birds and more concerned about their well-being, all the better.

Each chapter begins with a personalized experience that anyone visiting one of these national parks might have. At Olympic, for example, a walk up Hurricane Hill and a view of the incredible scenery and wildlife provide the introductory setting. A morning walk along the Shadows Nature Trail at Longmire and a birding walk around Mazanita Lake introduce readers to Mount Rainier and Lassen Volcanic National Parks, respectively.

Each chapter then continues with a description of the national park itself, including the plant and animal communities that exist there, visitor facilities available within the park, interpretive activities, and how to write or telephone for additional park information. The chapter then returns to the bird life, describing common birds within several of the park's most popular and accessible areas. Each chapter ends with a summary of the park's bird life and a list of a few key species.

WHERE TO BEGIN

A visit to any of the national parks should begin with a stop at the park's visitor center or information station to obtain a park brochure and activity sheet. These will contain basic information about roads and trails, accommodations, campgrounds, picnic sites, and hiking routes; details of interpretive activities; descriptions of the park's key resources; and so on. The numerous sites mentioned in the book can best be located by using the area map in each park brochure.

Common bird names used throughout the book are taken from the most recent checklist of birds published by the American Ornithologists' Union (AOU) and supplements and used in the most current field guides.

The section titled References, beginning on page 25, includes all references used in the writing of this book. It is hoped that the reader will utilize those books, articles, and reports for continued study of birds and of the habitats that are so essential to their survival.

Enjoy!

ROLAND H. WAUER

Acknowledgments

This book would not have been possible without the kind assistance of several employees of the National Park Service: superintendents, rangers, naturalists, resource specialists, scientists, and a few others. I especially want to thank the following individuals for their assistance in the field and for their reviews of pertinent chapters: Mac Brock, Dave Morris, and Kent Taylor at Crater Lake; Gil Blinn, Alan Denniston, Scott Isaacson, Mike Lafkos, and Steve Zachary at Lassen Volcanic; Chuck Barat and Michele Moore at Lava Beds; Gary Ahlstrand, Bill Dengler, Allan Grenon, and Kathleen Johnson at Mount Rainier; Cindy and Johnathan Bjorlund, Roger Christophersen, Bob Kuntz, and Terri Spencer at North Cascades; Roger Huffmann and Bruce Moorhead at Olympic; and Craig Ackerman, Sheri Forbes, and John Roth at Oregon Caves.

I also want to thank Jan Hartke and President John Hoyt of the Humane Society of the United States (HSUS) and Earthkind, respectively. The monetary assistance provided by HSUS made possible the travel and research required for this book. The contributions of both organizations are most appreciated.

Artist Mimi Hoppe Wolf's wonderful pen-and-ink sketches have greatly enhanced descriptions of many of the highlighted bird species.

Finally, I want to thank my wife, Betty, who accompanied me on all of my travels as I prepared for the writing of this book. I also thank her for her assistance with editing and for her patience, fortitude, and love.

Birds

of the Northwestern

National Parks

Fig. 1. Bald Eagle

BIRDS   *What They Are and How to Find Them*

The bond between birds and man is older than recorded history. Birds have always been an integral part of human culture, a symbol of the affinity between mankind and the rest of the natural world, in religion, in folklore, in magic, in art—from early cave paintings to the albatross that haunted Coleridge's Ancient Mariner. Scientists today recognize them as sure indicators of the health of the environment. And as modern field guides make identification easier, millions of laymen watch them just for the joy of it.
PAUL BROOKS

How often I have wished I could fly: to soar high over the mountains and valleys; to explore secluded places that are impossible to reach any other way; to escape this earthbound existence with the ease of a bird. These were among my secret desires as a youngster. How I envied the hawks and the swallows and even the tiny hummingbirds. They were the masters of my universe.

Of all the warm-blooded creatures, only birds and bats can fly for more than a few yards. Birds alone possess the combination of feathers, powerful wings, hollow bones, a remarkable respiratory system, and a large, strong heart. Marvelous flight feathers allow a bird to cruise at speeds of 20 to 40 miles per hour while flying nonstop across the Gulf of Mexico or the Arctic tundra. The tiny hummingbird has been clocked at 50 miles per hour. And the powerful Peregrine Falcon is thought to stoop at speeds in excess of 100 miles per hour.

A Blue-winged Teal banded in Quebec was killed by a hunter less than four weeks later in Guayana, a distance of more than 2,500 miles. A Manx Shearwater, taken from its burrow on Skokholm Island, Wales, and carried by airplane to Boston, returned to its burrow on the thirteenth day, having flown 3,000 miles across the Atlantic Ocean. And a Lesser Yellowlegs banded in Massachusetts was captured six days later 1,900 miles away on the island of Martinique, in the Lesser Antilles. That bird had averaged 317 miles per day.

Migrating birds usually fly at elevations of less than 3,000 feet, but observers at 14,000 feet in the Himalayas reported storks and cranes flying so high overhead, at an estimated elevation of 20,000 feet, that they could barely be seen through binoculars.

Other marvelous features of birds are their bill shapes and sizes. Anyone who has watched birds for any time at all cannot help noticing the diversity of feeding methods. Hummingbirds, for example, have long, thin bills especially adapted for probing deeply into tubular flowers, where the bird feeds on nectar. Many shorebirds, such as dowitchers and the Common Snipe, also have long bills, but they are much heavier for probing for food in mud. The Long-billed Curlew can reach into deep burrows to extract its prey.

The many insect feeders have dainty bills for capturing tiny insects. Vireos and warblers are gleaners that forage on trees and shrubs, picking insects off leaves and bark. A careful examination of feeding warblers will further suggest the size of their preferred food based on their bill size. Flycatchers' and swallows' bills are wider to enhance their ability to capture flies in midair. Woodpeckers' bills are specialized for drilling into insect-infested trees and shrubs to retrieve larvae.

Finches' bills are short and stout, most useful for cracking seeds or crushing armored insects. Crossbills are able to extract seeds from conifer cones. And grosbeaks are able to feed on much larger fruit, actually stripping away the husk from fleshy seeds. Many birds feed on fruit when it is abundant in late summer and fall and on insects at other times of the year.

Finally there are the predatory birds, with their variety of bill shapes and sizes. Raptors possess short, stout bills with a specialized hook used for tearing apart their prey. Wading birds have large, heavy bills for capturing their prey. And the bills of diving birds are hooked for catching fish and serrated on the edges for a better grip.

Feet are another fascinating feature of anatomy helpful for understanding a bird's requirements. Webbed feet suggest an adaptation to water for swimming, and flattened toes help birds walk on soft mud. Tiny, flexible toes suggest an ability to perch on small twigs and branches. And large, powerful feet with sharp talons are required to capture and grip prey.

There are approximately 9,600 kinds of birds in the world, about 900 of which are found in North America. Every type of bird has slightly different characteristics that permit it to utilize a slightly different niche (an ecological role based on the combination of its needs) than that of any other species. Whenever two or more species have the same needs, in all likelihood only one will survive.

A bird is a very specialized creature, but its bill and feet are usually less obvious than its plumage, the sum total of its feathers. A bird's plumage is unquestionably its most obvious and usually most attractive characteristic. This is especially true for the more colorful and contrasting birds, such as warblers, hummingbirds, some waterfowl, and some finches. Birds are the most colorful of all vertebrates.

Feathers reveal every color in the rainbow. The colors we see are the product of either pigments or the reflection and refraction of light due to feather structure. The concentration of pigments produces intensities of color, such as the vivid red of a male Vermilion Flycatcher and the diluted red of a female Northern Cardinal. Total lack of pigment production results in white plumage. Other colors are the result of light that is reflected or absorbed by feathers. The bright blues of Steller's Jays and bluebirds are due to a particular arrangement of cells in the feather, which produces iridescence. A dull velvet color is the opposite of iridescence.

Of all the aesthetically pleasing characteristics of birds, birdsong may be the most enduring. Louis Halle wrote, "As music is the purest form of expression, so it seems to me that the singing of birds is the purest form for the expression of natural beauty and goodness in the larger sense, the least susceptible of explanation on ulterior practical grounds."

But birds possess additional values that are sometimes ignored or taken for granted. For instance, certain birds are extremely adept at catching and consuming large quantities of insects, many of which are considered pests. These include obnoxious insects as well as those that are a serious threat to various crops.

Human beings have used birds from earliest history. Birds were worshipped by many early civilizations. Cormorants were ringed for catching fish. Pigeons carried our messages. Songbirds were taken into mines and brightened our homes with their wonderful songs. In literature, Samuel Coleridge has immortalized the albatross, Percy Shelley the lark, and Edgar Allan Poe the raven. The concept and development of manned flight was derived from our observations of birds. Every state has an official bird, many of which highlight flags and seals. Postage stamps often display common bird species. And the most powerful country in the world employs a bird as its symbol: America's Bald Eagle is one of the most visible symbols in the United States.

Birds truly are an intricate part of the human ecosystem, an important link to nature. Birds, more than any other creatures, are obvious and omnipresent companions to the human community.

BIRDING FOR FUN

There comes a time when those of us with a natural curiosity and appreciation for the outdoors want to know the names of the various creatures we see around us. The spark to identify birds may be kindled by some exceptional happening or a special sighting. Watching a family of Steller's Jays at a campground as they actively investigate you and your food supply, or suddenly being mobbed by a flock of Cliff Swallows at a nest site, is likely to foster interest in those species.

But identifying those birds can be somewhat difficult unless you know where to begin. Although the average park visitor usually can identify more birds than he or she might first assume, further bird identification requires some basics, just like any other endeavor. The basics include two essential pieces of equipment: a field guide and a pair of binoculars.

Several very good field guides are available that use the bird identification technique developed by Roger Tory Peterson. Peterson's field guides and those published by the National Geographic Society, Golden Press, and American Bird Conservancy use bird paintings.

These guides are preferred over those with photographs, because the paintings highlight key features that only occasionally are obvious in photographs.

Binoculars are absolutely essential for identifying, watching, and enjoying most birds. Binoculars vary in power, illumination, and field of vision, as well as price. The most popular birding binocular is an 8 x 35 glass. Eight is the power or magnification: an 8 power (or 8x) binocular magnifies a bird eight times, a 7 power (or 7x) binocular seven times, and so on. Thirty-five is the diameter of the objective lens in millimeters and is used to illustrate illumination. Illumination (brightness) can be determined by dividing the magnification into the size of the optical lens; a 7 x 50 binocular produces a brighter image than an 8 x 35 binocular by 7.1 to 4.4. The larger 50 mm binocular provides a brighter image than the 35 mm binocular and is better in dim light; however, it usually is too heavy for a full day in the field. Also, binoculars with a power of 9x or more are often too powerful for beginners who are not yet comfortable with holding binoculars perfectly still. Pocket-sized, lightweight binoculars (those with a small objective lens) are good for occasional use, but continuous use can cause eyestrain. Select binoculars that are best suited for you.

Field of vision, determined by the binocular design, usually is also marked in degrees (angle visible out of 360°) or feet (width visible out of 1,000 ft. or 305 m). The 7.3-degree field of vision (out of 360°), for my wife's 9 x 35 Discoverer binoculars, and the 395 feet of vision (out of 1,000 ft.), for my old 9 x 35 Burton binoculars, will remain the same no matter how far away we are from the bird being observed.

In addition, central-focus binoculars are a must. Minimum focusing distance is important as well for focusing on a bird that may be as close as 12 to 15 feet (4–6 m) away. Binoculars range widely in price, but the moderately priced ones usually work just as well as the most expensive, which may be more water resistant, less inclined to fog, and armored for rough use.

Using binoculars usually requires some experimentation, but the skill is easy to learn. First make sure that the right ocular is set at 0 for 20/20 vision. Then while looking directly at an object, bring the binoculars up into position without changing your position or looking elsewhere, and use the center wheel to focus on the object. A few tries will produce immediate success.

The next step is to get acquainted with your field guide. Start by leafing through the entire guide and locating the first page of tyrant flycatchers, just after woodpeckers. Flycatchers and all the birds illustrated on following pages are perching birds (songbirds). All the nonperching birds (e.g., seabirds, waders, waterfowl, raptors, shorebirds, gulls and terns, grouse, hummingbirds, woodpeckers) are located in the first portion of the book.

Next read the introductory section, especially the discussion about field marks. You will find a drawing of a typical bird showing basic field marks. Look these over so that you have a good idea of where the bird's crown, eye line, eye ring, chin, upper and lower mandibles, flank, uppertail and undertail coverts, wrist, wing bar, and so on, occur. Be ready to refer back to this illustration for help when necessary.

Now that you have discovered the value of a field guide, it is time to start identifying real birds. You should have an idea of what features to look for. The following suggestions provide an identification strategy of sorts:

1. Size. It is a good idea to relate bird size to those species you already know. For instance, consider five categories: sparrow size, robin size, pigeon size, duck size, and heron size. With a few exceptions, such as the Common Raven, any bird the size of a duck or larger is a nonperching bird and will be found in the first half of the field guide. By thinking in terms of size, you immediately know where to start your search. Also, one can often pick out odd-sized birds in a flock for further attention or to recognize different species that might be foraging together. For example, a tiny bird within a party of warblers will more than likely be a chickadee, kinglet, or Brown Creeper.

2. Shape and behavior. Does your bird possess any outstanding features? Is it a wader with long legs and an upright posture? Possibly a heron. Is it walking along the shoreline? Possibly a shorebird. Is it swimming on a lake or river? Probably a waterfowl or gull. Is it soar-

ing high in the sky? Possibly a Turkey Vulture or hawk. Is it perched on a wire or tree limb? Certainly a perching bird. Is it a perching bird eating seeds at a feeder? Probably a sparrow or finch. If it is smaller than a sparrow, is creeping up a tree trunk, and is all brown, it is sure to be a Brown Creeper.

3. Color and pattern. Many birds possess an obvious plumage that is an immediate giveaway. Cardinals, crows, robins, Yellow Warblers, and Red-winged and Yellow-headed Blackbirds are the first to come to mind. Their bold and obvious colors or patterns, or both, stand out like a sore thumb. But many of their neighbors will require a little more study. Do the all-white underparts extend onto the back, or does your bird have only white wing bars? Does its white neck extend only to the lower mandible or onto the face? Does its reddish color extend onto the back, or is it limited to the tail and wings? Do the yellow underparts include the throat and belly or only the chest? Answering these questions will eventually become like second nature.

FIELD TECHNIQUES

Bird-finding techniques are often personal ones, and you will discover your own methods. For example, I like to move very slowly through a particular habitat, trying to discover all of the birds within the immediate area. I find that part of birding most enjoyable and challenging. Other birders prefer to move faster, stopping only to watch birds that become obvious. This method is based on the concept that they will find more birds by covering more ground. That is definitely the reason for visiting as many habitats as possible, but I believe that the largest number of species can be found by slowly moving through each habitat, making yourself part of the scene, both physically and mentally.

There are definite clues to bird-finding that you can use to your advantage. First are bird sounds. During the breeding season, birdsong is the very best indicator of a bird's presence and location. Songbirds often sing throughout the day. They almost always sing at dawn and dusk, but a few species sing only at dawn. The more serious birders get out at dawn to experience the dawn chorus while other birders are still asleep. Most of these birds, however, can usually be found throughout the day.

Rustling leaves in the underbrush can be another valuable clue. Leaf rustling can be caused by numerous creatures, but when the leaves seem to be thrown back as if being cleared away for finding food underneath, the rustler is likely to be a thrasher, Fox-Sparrow, or towhee.

Songbirds tend to ignore intruders who are quiet and move slowly, unless they get too close to a nest or fledgling. You can get surprisingly close to songbirds by moving slowly and not making any sudden motions. Also, wearing dull clothing instead of bright and contrasting clothing helps you to blend into the bird's environment, usually permitting closer viewing.

Some of the nonperching birds will permit a slow, cautious approach, but the wading birds, ducks, and raptors are not as trusting. You will need to observe these birds from a distance, and you may want to use a spotting scope for your observations. Or you may be able to use a blind, sometimes installed at bird-viewing sites.

During the nonbreeding portion of the year, birds often occur in flocks or in parties. Flocks of waterfowl or blackbirds can number in the hundreds or thousands and be readily visible from a considerable distance. But identifying all the members of a party of songbirds moving through the forest will require quiet study. It is possible to wander through the woods for some time before discovering a party of birds that may include a dozen or more species. Migrant songbirds usually travel in parties that can include hundreds of individuals of two or three dozen species. If you find such a party, remain still and let the party continue its feeding activities without disturbing it.

In cases in which a bird party is just beyond good viewing distance, you can sometimes attract a few of the closer individuals by "spishing"—making low, scratchy sounds with your teeth together and mouth slightly open—a few times; attracting the closer individuals often entices the whole flock to move in your direction. However, I find that spishing within a bird party tends to frighten some species off or to move the party along faster than it might otherwise go.

At times a bird party is concentrated at a choice feeding site, such as near flowering or fruiting trees and shrubs. So long as they are not frightened or unduly agitated by noises or movement, they may remain and continue feeding for some time, once they overcome their initial concern regarding your presence. Also, their activities will tend to attract other birds, allowing you to see a broad spectrum of birds at one spot.

Generally, birding along a forest edge, often along the edge of a parking lot, can produce excellent results in the early morning. Bird parties prefer sunny areas at that time of day, to take advantage of greater insect activity. Within two or three hours, however, feeding birds tend to move into the cooler vegetation, especially on hot, sunny days.

Birds may then need to be enticed into the open; many species respond well to some sounds. Spishing often works very well. Squeaking sounds made with your lips against the back of your hand or finger may work at other times. Birds are naturally curious and will often come to investigate. At other times, spishing or squeaking seems to frighten birds away. And some species will be attracted once but difficult to fool twice. So always be prepared to focus your binoculars on a bird immediately when it pops up from a shrub or out of a thicket.

As mentioned above, the best way to find a large number of birds is to visit a variety of bird habitats. All birds occur in preferred habitats, especially during their nesting season. But they tend to frequent a broader range of sites in migration and in winter. Learn where species can most likely be expected. For instance, Boreal Chickadees occur only in northern coniferous forests; this species cannot be found at Crater Lake National Park. And one cannot expect to find a roadrunner in the boreal forest. New birders should learn to take advantage of the range map and habitat description for each species that is included in the field guides. This can save time and considerable embarrassment.

Birding by song is often left to the experienced birder, but many novices are as well equipped to use birdsongs as many of the experts. For anyone with an ear for melody, many records, tapes, and CDs are available to help you learn the birdsongs. During spring and summer, there is no better method of bird identification. When tiny passerines are singing from the upper canopy of the forest, finding and observing those individuals can be most difficult. But identifying their songs is an instant method of recognition that does not produce eye and neck strain from staring into the high canopy for hours on end. And observing rails can also be trying, if not outright dangerous. Fortunately, rails and other marsh birds also sing their own unique songs that can usually be easily identified.

Much of the knowledge required to make quick bird identifications must come from field experience. An excellent shortcut is spending time with an experienced birder who is willing to share his or her knowledge. That person can pass on tidbits of information that otherwise might take years to acquire. Most national parks have staff naturalists who give bird talks and walks during the visitor season. This kind of assistance can be extremely worthwhile for bird-finding and identification.

BIRDING ETHICS

As with any other activity, there are certain rules of the game. Birding should be fun and fulfilling. It can be a challenge equal to any other outdoor endeavor, but it should never become so all-consuming as to threaten the bird's health and habitat. Any time we are in the field, we must realize that we are only visitors to that habitat on which a number of birds depend for their existence. We must not interfere with their way of life. Disturbing nests and nestlings, for whatever reason, cannot be tolerated. Tree-whacking, to entice woodpeckers and owls to peek outside, is not acceptable.

Most national parks are adequately posted, but sometimes just plain thoughtlessness can lead to severe impacts on the environment. These acts range from short-cutting to actually driving over a tundra or meadow. Respect closures in the park; they are there for a very good reason. The survival of nesting terns or peregrines may depend on such measures.

The hobby of birding can be a most enjoyable pastime. It is one that costs very little and can be done with little or no special training. It can be pursued alone or in a group and at any time of the day or night. And there is nowhere on earth where birds are not the most obvious part of the natural environment.

Early naturalist Frank Chapman, in his *Handbook of Eastern North American Birds*, summarizes the enjoyment of birds better than anyone else. Chapman writes that birds "not only make life upon the globe possible, but they may add immeasurably to our enjoyment of it. Where in all animate nature shall we find so marvelous a combination of beauty of form and color, of grace and power of motion, of musical ability and intelligence, to delight our eyes, charm our ears and appeal to our imagination."

Fig. 2. Peregrine Falcon

PARKS AS ISLANDS

All living things possess an intrinsic value which is beyond calculation. Humans as rational beings are responsible for safeguarding forms of life which we did not create but suddenly have the power to destroy. By knowingly causing extinction of these species and their habitats, we sacrifice a part of our humanity. PHILLIP M. HOOSE, 1981

The last viable Peregrine Falcon populations anywhere in North America south of Alaska were those remaining in national parks in the Rocky Mountains, Colorado Plateau, and West Texas. The discovery that populations of these and several other high-level predators were being decimated by DDT and other chlorinated hydrocarbons, and the eventual banning of DDT use in the United States and Canada in 1972, came too late to save any of the eastern Peregrines. The last active aerie in the Appalachian Mountains was at Great Smoky Mountains National Park. The entire population of that subspecies became extinct in three decades. Fewer than thirty pairs of eastern Peregrines were known in the United States by 1975. Peregrines in Big Bend, Black Canyon, Dinosaur, Grand Canyon, Mesa Verde, and Zion (all national park units), however, were well enough isolated and in sufficient numbers to ensure the survival of an adequate breeding population.

These examples demonstrate the value of national parks as natural refuges. The western parks provided last strongholds in which Peregrine populations could withstand human-induced pollutants. In most cases, those Peregrines fed primarily on resident birds that had not been subjected to DDT elsewhere. But in the case of the Great

Smoky Mountains population, insufficient buffers existed, and the eastern Peregrine disappeared forever.

During the 1980s, when Peregrine restoration programs were being implemented, park sites in the Great Smoky Mountains and at Isle Royale, Yosemite, Sequoia, and Zion were among the first selected. Nearly 2,500 Peregrine Falcons were released in the West, according to James Enderson, leader of the Western Peregrine Recovery Team. By 1990, Peregrines once again began to frequent their old haunts, including the actual nesting of pairs at several areas. The finding of fifty-eight active aeries at Grand Canyon in 1989 suggests that Peregrine populations have recovered sufficiently to consider delisting the species. The locations of current populations further highlight the importance of national parks to species' recovery.

Despite an apparent Peregrine "fix," many other bird populations continue to decline. The most serious losses are occurring in Neotropical species, long-distance migrants that nest in the United States and Canada and winter in the tropics, in the Greater Antilles, Mexico, Central America, and to a lesser extent in northern South America. According to U.S. Fish and Wildlife Service Breeding Bird Survey data, forty-four of seventy-two Neotropical species declined from 1978 to 1987. These include almost all the warblers; five vireos; five flycatchers; and various thrushes, buntings, orioles, tanagers, cuckoos, grosbeaks; and the Blue-gray Gnatcatcher.

The reasons for the declines are varied. Neotropical migrants are less adaptable than most resident species. They have a shorter nesting season, with only enough time to produce one brood before they must depart on their southward journeys. Long-distance migrants tend to arrive on their breeding grounds later and depart earlier. They also produce smaller clutches than the full-time residents. And most of the Neotropical species place their nests in the open, either on the ground or on shrubs or trees. Their nests, therefore, are more susceptible to predators and brood parasitism by cowbirds than those of the full-time residents, many of which are cavity nesters (woodpeckers, chickadees, titmice, wrens, and bluebirds). If a raccoon, skunk, or fox destroys the nest of a full-time resident, the bird could start over, but one episode of predation or parasitism can cancel an entire breeding season for a Neotropical bird.

Breeding bird studies in the fragmented environment of Rock Creek Park, Washington, D.C., conducted from 1947 through 1978, revealed that six Neotropical species (the Yellow-billed Cuckoo, Yellow-throated Vireo, and Parula, Black-and-white, Hooded, and Kentucky Warblers) could no longer be found to nest. And several other species, including the Acadian Flycatcher, Wood Thrush, Red-eyed Vireo, Ovenbird, and Scarlet Tanager, had declined by 50 percent. Conversely, at Great Smoky Mountains National Park, with its 494,000 acres of mature forest, breeding bird censuses conducted in the late 1940s and repeated in 1982 and 1983 "revealed no evidence of a widespread decline in Neotropical migrants within the large, relatively unfragmented forest" of the park, according to the National Fish and Wildlife Foundation.

These divergent examples, peregrines in Big Bend, Grand Canyon, and other western parks, and Neotropical breeders in the Great Smoky Mountains, demonstrate the value of large natural parks as preserves for the perpetuation of wildlife resources.

THREATS TO THE PARKS

North America's national parks are not immune, however, to the abundant environmental threats facing wildlife. Every park has experienced impacts that threaten its ecological integrity. Although its exterior shell may appear unchanged, and the average visitor may find the scenery looks pretty much the same from year to year, a number of strands in the parks' fragile ecological webs have been damaged.

During the early years, most of the natural parks had sufficient buffers to insulate them from development and pollution outside their borders. But with continued population growth and increased adjacent land uses, the parks' buffer zones have dwindled. Many parks are bordered by farmlands that are maintained by chemicals, forests that are clear-cut, and increasing numbers of industrial centers, malls, and housing developments. Widespread air pollution reaches great distances and affects even the most remote parkscapes.

Long-term monitoring of air quality values in several of the western parks reveals that prevailing winds, especially in summer, carry pollutants from far-away urban and industrial areas. Plant and animal communities all over the globe are linked by the air that is moved around by weather patterns. In a sense, our world is like a large room that shares the same recirculated air. This circulation has created pollution in over one-third of the national parks. Even isolated locations are affected. At pristine Isle Royale, wind-borne chemicals have turned up in inland lakes; at Big Bend, visibility is only half of what it was a decade ago.

Inside the parks, roadways, trails, campgrounds, and other facilities permit greater human use of the resources. Often these features are poorly sited and designed, thus increasing fragmentation and stressing resources already threatened by external perturbations.

Habitat degradation within the parks due to improper management also can have serious consequences for the park's bird life. Any fragmentation reduces the integrity of the unit, lowering its value for wild species. New developments increase access to the forest interior for predators that feed on birds and their eggs; parasitic cowbirds that lay their eggs in other species' nests; exotic House Sparrows, European Starlings, and other invaders that compete for nesting space and food; and exotic plants that can drastically change the habitat.

A number of recent studies suggest that cowbird parasitism can affect even songbirds in large forest tracts and may be the major cause of the decline of many Neotropical migrants. Researchers have concluded that cowbirds "will commute up to seven kilometers [4.35 mi.] from feeding areas to search for nests to parasitize." John Terborgh reports in the May 1992 issue of *Scientific American* that "A seven-kilometer [4.35 mi.] radius describes a circle of 150 square kilometers [58 sq. mi.], equal to 15,000 hectares [37,065 acres]. It is disturbing to think a forest that might offer at its center a haven from cowbird parasitism would have to be at least that size."

Cuts into the forest interior also increase populations of other open-area birds, such as American Crows, jays, magpies, and grackles, which prey on other birds and their eggs and hatchlings.

Once a park's natural ecosystem has been damaged by fragmentation and pollution, all the resources become more susceptible to impacts from natural disasters, such as hurricanes, floods, fires, and diseases. These catastrophes can seriously affect small-bird populations that already have been reduced by pollution, predators, parasites, and competitors.

WILDLAND FIRES AND THEIR EFFECTS ON BIRD LIFE

Environmental changes are part of every natural system. But a healthy bird population is better able to withstand those changes than one that has been weakened or diminished by other factors. Wildland fires, which occur in most forest, shrub, and grassland communities, are one example of such changes. Indeed, many plants and animals are fire dependent. Some pine cones must burn to open, drop their seeds, and regenerate. Woodpeckers frequent freshly burned sites to feed on various wood-boring beetles that are attracted to trees weakened by natural fires. Many raptors are attracted to prairie fires to feed off the displaced rodents and insects.

Wildland fires have received considerable scrutiny by the public and government officials since the highly visible Yellowstone National Park fire of 1988. Almost one million acres of Yellowstone's parklands were affected by that burn. Although park officials readily point out the negative effects of fire, largely in areas where excessive fuel loads have built up over too many years without burning, they also are eager to discuss the benefits of fire to the Yellowstone ecosystem. Fire is a natural part of the ecosystem. Old-age forests do not have the diversity of wildlife that occurs in mixed-aged forests created by fires. Fire opens the forest so that new vegetation, such as grasses, aspens, and a variety of shrubs that had been overcome by the old-age forest can contribute to the mix of habitats. Terry Rich's article, "Forests, Fire and the Future," published in *Birder's World* (1989), includes a good summary of the Yellowstone fire and its effects on the park's bird life.

Based on my own research at Bandelier National Monument, the 1977 La Mesa fire initiated a series of changes that continue twenty years later. All three of my study sites revealed increased numbers of bird species and populations following the fire. Significant population increases occurred almost immediately in woodpeckers, with minor increases in all the other insect feeders, such as Violet-green Swallows, nuthatches, and warblers. Seed feeders, such as sparrows, juncos, and finches, initially declined in varying degrees but soon increased with the newly available grass seed. Once woodpeckers became established, other cavity nesters such as Ash-throated Flycatchers, Violet-green Swallows, Mountain Chickadees, nuthatches, and bluebirds increased to take advantage of vacated nest holes. Predators also increased with the additional prey base and more open character of the landscape. A few lowland species, such as Mourning Doves, Ash-throated Flycatchers, Say's Phoebes, Western Scrub-Jays, and Spotted Towhees, moved in to utilize the open and warmer terrain. Snag-fall provided increased habitats for House Wrens. Although the initial influx of Downy, Hairy, and Three-toed Woodpeckers and Northern Flickers tapered off by the fifth year, Lewis's Woodpeckers moved into some of the vacated nest holes. And vireos, Virginia's Warblers, Black-headed Grosbeaks, towhees, and Dark-eyed Juncos were soon able to take advantage of the young aspens and oak thickets.

THE FUTURE

Although extinction is part of the natural process, the rate of extinction has never been as swift as it is at present. The International Union for the Conservation of Nature and Natural Resources (IUCN) predicts that by the year 2000 the world will have lost 20 percent of all extant species.

The greatest losses are occurring within the tropical forests, where many of our songbirds spend their winters. The Council on Environmental Quality's *Global 2000 Report to the President* states: "Between half a million and 2 million species—15 to 20 percent of all species on earth—could be extinguished by 2000, mainly because of loss of habitat but also in part because of pollution. Extinction of species on this scale is without precedent in human history. . . . One-half to two-thirds of the extinctions projected to occur by 2000 will result from the clearing or degradation of tropical forests."

In North America, at least 480 kinds of plants and animals have become extinct since Europeans first arrived. Seven species have disappeared since 1973, when the U.S. Congress enacted the Endangered

Species Act. More than six hundred species have since been listed as "threatened" or "endangered." Endangered species are those in danger of becoming extinct; threatened species are those on the verge of becoming endangered. Howard Youth of the Worldwatch Institute reports that worldwide "about 1,000 bird species—more than 11 percent—are at risk of extinction, while about 70 percent, or 6,300 species, are in decline."

Today the concept of threatened and endangered species is an accepted part of our world. Significant decisions are based on whether or not a species is listed. And many of our "T and E" species have become household terms. Who has not heard of the plight of the Peregrine Falcon, Spotted Owl, and humpback whale?

The shortcoming of the Endangered Species Act is that it addressed individual species instead of communities of plants and animals. Attempts to restore species do not always give adequate attention to the natural processes on which they depend. And at a time of inadequate funding and moral support, only the more charismatic species receive attention.

An endangered ecosystem act would have much greater success in saving species by giving adequate protection to large tracts of intact landscape (at least 58 sq. mi.) that contain several threatened and endangered species. These larger areas are the essence of the national parks.

The bottom line is that our North American birds are losing their breeding grounds, winter habitats, and many of the stopover places in between.

WHAT IS BEING DONE WITHIN THE PARKS?

Much has been written about the threats to park resources. The National Park Service itself has been in the forefront of expressing concern about those threats. A major "State of the Parks" initiative was undertaken in 1980 and 1981 to identify the threats and to establish a program for preventing additional threats and mitigating current impacts. Parts of that strategy continue to be implemented, but other portions were eliminated, reduced, or ignored because of in-house bureaucracy or insufficient funding.

In addition, many of the national parks within the western states, from Alaska to the deserts of California and Arizona, are involved in one way or another with bird-oriented research, monitoring, and restoration activities. Pertinent examples include an abundance of Christmas Bird Counts, numerous breeding bird surveys, and several studies involving threatened or endangered species such as Bald Eagles at North Cascades, Olympic, and Lassen; Peregrine Falcons at Olympic and Mount Rainier; Marbled Murrelets at Olympic; and Spotted Owls at Olympic and Mount Rainier.

In addition, the National Park Service is participating in the Neotropical Migratory Bird Conservation Program, as coordinated by the National Fish and Wildlife Foundation. The primary focus of this program will be to "integrate research, monitoring, and management (including ecological restoration activities) on behalf of migratory nongame birds" for "the conservation of Neotropical migratory birds" (National Fish and Wildlife Foundation 1990).

Many of the national parks also are participating in "Partners in Flight," a consortium of government agencies and private organizations. This is an international program to identify and conserve species of birds and their habitats that are in peril.

Early emphasis has been placed on development of a network of parks and protected areas in the Western Hemisphere that are linked by Neotropical migratory birds. Migratory bird-watch programs have begun in more than three dozen parks as another way to develop linkages between pertinent park units.

Yet despite the parks' mandates for unaltered ecosystems, the dual charge of "protection and enjoyment" has too often been interpreted to mean that the parks are meant primarily for people instead of the resources they contain. And so even the largest of the parks have undergone changes to benefit the human visitor, while creating more chinks in the park's ecological integrity. Taken separately, these incremental bites may seem insignificant, but together they have eaten into the essential fabric that keeps parks viable. Many of our national parks are little more than skeletons of their former selves.

More and more, today's national parks are becoming islands within a great sea of disturbance. They can be equated with sea islands, with

Fig. 3. Marbled Murrelet

few connections to continental sources for species renewal. The longer an island is isolated, the less its flora and fauna have in common with other communities, and the greater the likelihood of species loss.

THE VALUE OF NATIONAL PARKS

How important are national parks for the perpetuation of our North American bird life? Except for a small number of the largest and most remote public forests and refuges and an insignificant scattering of private preserves, only the national parks are dedicated to the preservation of complete ecosystems. Most other managed areas are primarily dedicated either to the perpetuation of one or a few species, or to the area's multiple-use values. The perpetuation of unaltered ecosystems is too often given secondary importance.

Despite changes, the national parks still represent some of the finest of our natural environments. Many units contain some of the least disturbed habitats in North America. And each national park increases in value with every passing day.

The national parks contain far more than tracts of natural landscape, scenic beauty, and places of inspiration. The parks literally serve as biological baselines for the continent, where gene pools of diversity can be found that have disappeared almost everywhere else. The national parks contain our last remaining outdoor laboratories in which we can learn about the past in order to prepare for the future. They represent the last reasonably intact examples of what North America was like before our resources were exploited for the benefit of a few.

A survey of the national parks quickly reveals that the park system of the United States is far from complete. Many of North America's major biotic communities have been left unprotected. Examples include an extensive tall-grass system in the Great Plains and a tropical brush and riverine system in South Texas. The National Park Service has recognized this shortcoming and taken steps to identify the needs but must wait for congressional approval before continuing the process of actual designation and acquisition.

In addition, only a few of the parks and reserves are large enough to represent entire ecosystems. Few parks contain a complete spectrum of natural processes that shape the ecosystem, so that nutrient, hydrologic, fire, and other natural cycles are allowed to function unfettered by human constraints.

The key to long-term perpetuation of native species is the complete protection of intact ecosystems. Their borders must be established to include entire watersheds and all essential habitats for larger wildlife. They must be fully protected from all degrading activities, including grazing, timber-cutting, mining, all forms of pollution, fragmentation by overdevelopment or overprotection (such as construction of fire roads), and various forms of recreation incompatible with the purpose of the park.

C. F. Brockman understood the greatest value of the parks when he wrote:

> The national parks are charged with the obligation of preserving superlative natural regions, including wilderness areas, for the benefit of posterity. Attentiveness to the pleasure and comfort of the people is essential

but it cannot mean catering to absolutely unlimited numbers unless the second function is to destroy the first. In a theatre, when the seats in the house have been sold out and the available standing room also has been preempted, the management does not jeopardize the main event by allowing still more onlookers to crowd upon the stage and impede the unfolding of the drama.

THE NORTHWESTERN NATIONAL PARKS

I think we all agree that a national park is not merely scenery. A national park embodies something that cannot be found everywhere—it embodies history, a way of life, primitive experience, early environment. It has the elements capable of providing that lifting of the spirit for which modern civilization is willing to pay so much. A national park is specifically dedicated to these intangible and imponderable qualities. OLAUS J. MURIE

The national parks in the northwestern corner of America contain some of the most beautiful, most rugged, most regal, and most inspiring scenery and some of the most diverse and fascinating flora and fauna to be found anywhere in North America. The best of all this can be found within seven representative natural park units in the states of Washington and Oregon and in Northern California: Olympic, North Cascades, Mount Rainier, Crater Lake, and Lassen Volcanic National Parks and Oregon Caves and Lava Beds National Monuments.

The birds and their essential habitats within these seven areas also are some of the most fascinating and unusual to be found anywhere. More than four hundred species occur from the rugged shoreline of the Olympic Peninsula to the rocky crags of the North Cascades, and to the deep blue caldera waters of Crater Lake and the sagebrush flats of Lava Beds.

The following seven chapters provide the novice bird-watcher, as well as the avid birder, with fresh insights into the bird life of these amazing areas. You are invited to join me on a birding excursion through America's key northwestern parks.

Olympic National Park, Washington

The snow-capped Olympic Mountains stretched out before us like a string of bright jewels in the morning light. The deep green forest below made the mountains stand out in bold relief. Behind us, to the north of Hurricane Hill, the steep slopes descended toward the deep blue waters of the Strait of Juan de Fuca. Vancouver Island formed a low green ribbon beyond.

All around us were subalpine meadows, only recently freed from snow, with scattered stands of subalpine fir. Bruce Moorhead and I had walked the paved 1.5-mile Hurricane Hill Trail that mid-June morning, admiring the grand beauty and watching for wildlife along the way. Black-tailed deer were plentiful. Bruce pointed out a black bear crossing a steep slope at about the midway point. And in a high meadow near the summit, several Olympic marmots stood guard at their burrows. A pair of youngsters were standing on their back legs pushing each other back and forth in friendly jousting.

We encountered a number of birds along the trail as well. A pair of Gray Jays inspected us at the parking area but sailed silently away through the forest when they discovered we were not picnickers with handouts. The musical trills of Dark-eyed Juncos were all about us. One of these little black-hooded birds, with obvious white-edged tail feathers, was feeding on the ground, hopping along in its search for seeds and insects. The cheerful caroling of American Robins and lively songs of Yellow-rumped Warblers were also commonplace around the parking area that morning.

One brightly marked Yellow-rumped Warbler, sitting at the top of a low fir tree, provided us with a marvelous view. Its coal black face and underparts contrasted with its bright yellow throat, cap, sides, and rump. A truly gorgeous creature! Its song consisted of a series of sharp, slightly rising "deedle" notes, followed by a falling "che-che-che" at the end. One of Olympic's most common songbirds, Yellow-rumpeds are identified by their black chests and five yellow spots: on their caps, throats, sides, and rumps. This bird was known as "Audubon's Warbler" until it was lumped with the eastern "Myrtle Warbler"; see the chapter on Lassen for details.

American Robins were also numerous. I watched one of these red-breasted birds gathering nesting materials at the edge of the parking area. It had accumulated a huge billful of lichens, mosses, and grasses. Melting snow, still present in the deep shade, would also provide the required mud for nest construction. Another individual was searching for food, running forward several feet, then stopping to search the ground for earthworms, spiders, and insects. Between each spurt it stood upright, alert for any dangers that might be present. It watched us carefully as we passed by.

Several other birds were detected as we walked along the trail. Golden-crowned Kinglets sang high-pitched double notes from stands of firs. Northern Flickers called loud "kleer" notes; Townsend's Solitaires called metallic "eek" notes on the slope to the west; a pair of Violet-green Swallows passed overhead; nasal "ank-ank-ank" notes of Red-breasted Nuthatches rang from the forest below the trail; a Hairy Woodpecker called there too; several Hermit Thrush songs resounded from the forest; a Blue Grouse flew up into a fir tree from a distant lush meadow; and a pair of Pine Siskins flew off a nearby perch, calling buzzy "shreeee" notes as they disappeared into the forest.

We topped off into an open meadow, edged with fir and filled with the tinkling sounds of **Horned Larks.** We located one individual sitting on a slight rise, alert to our presence, with its pair of black "horns" raised over its yellow, black, and brown head. The horns are the raised ends of black feathers that emanate from the forehead in a

V pattern. Suddenly, while we were watching, it flew up at a steep angle, rising several hundred feet in the air, where it began to circle. We had to use binoculars to see it well. Every four to eight seconds it glided, spread its black tail, displaying the obvious white edges, and sang its territorial song: "thin and unmusical, suggesting the syllables 'tsip, tsip, tsee-di-di,'" according to Ralph Hoffmann in *Birds of the Pacific States*. For about five minutes it circled an area we estimated at about 2 acres, singing its distinct song from on high. And then, without a hint, it plunged rapidly back to earth, wings folded in a power dive, and landed very close to where we had first found it. A wonderful avian encounter on a glorious day in the Olympic highlands.

THE PARK ENVIRONMENT

Olympic National Park is a land of superlatives. Not only does it contain magnificent scenery, but it also has one of the largest and best examples of virgin temperate rain forest in the Western Hemisphere, the

Fig. 4. Horned Lark

largest intact stand of mixed conifer forest in the Pacific Northwest, the longest stretch of wild coastline remaining in the contiguous states, and the largest truly wild herd of Roosevelt elk. Further, of the park's more than 1,200 plants, 237 birds, and 70 land and marine mammals, at least eight kinds of plants and sixteen kinds of animals are endemic to the Olympic Peninsula, found nowhere else.

It was this combination of features that prompted the United Nations Educational, Scientific, and Cultural Organization (UNESCO) to recognize Olympic as a Biosphere Reserve in 1976 and as a World Heritage Site in 1981. In 1988, Congress designated more than 95 percent of Olympic's approximately 900,000 acres as a wilderness area. Several park rivers were given "Wild and Scenic River" status in 1991. And in 1994, the coastal section of the park became a significant part of the greater Olympic Coast National Marine Sanctuary.

Olympic National Park contains an extremely wide diversity of environments, each worthy of national park status on its own. The 57-mile-long coastal strip, between Shi Shi Beach and Kalaloch, includes magnificent ocean views, bays, islands, seastacks, cliffs, headlands, and beaches. More than 350,000 acres of temperate, old-growth rain forest are found in the Hoh, Queets, and Quinault Valleys. And the higher slopes, ridges, and peaks reach 7,965 feet at the summit of Mount Olympus. More than 123,700 acres lie above the tree line, where annual snowfall supports about sixty glaciers and numerous perpetual snowfields.

Olympic's vegetation was studied and classified into six rather distinct communities by Jerry Franklin and C. T. Dryness (1973). The Sitka spruce zone is a lush rain forest community along west-side rivers and near the ocean, "where there is no effective summer drought." It is dominated by Sitka spruce, western hemlock, and western red cedar. Ferns and a huge variety of epiphytes are abundant here. Douglas-fir is confined to better-drained soils of river terraces, valley walls, and where wildfire or wind disturbance has occurred above the Sitka spruce zone. The western hemlock zone is more extensive and occurs throughout the lower elevations, "where summer droughts can be pronounced." This community is characterized by Douglas-fir, western hemlock, and either western red cedar or grand fir. At mid-elevations, especially on the western slopes, is a Pacific silver fir community, with mountain hemlock as a subdominant species, "where soils are colder and winter precipitation often falls as snow." Above the silver fir and western hemlock zones is the subalpine forest, or high-elevation forest zone, "where winter precipitation falls mainly as snow." Dominant trees include mountain hemlock and subalpine fir in wetter areas and subalpine fir, lodgepole, and whitebark pines in dry areas. Above the tree line, at an elevation of about 2,000 feet, is the alpine zone, where much of the landscape looks like a miniature rock garden in summer.

The National Park Service operates visitor centers and information stations at a number of locations. The Pioneer Memorial Museum and Visitor Center is located at the edge of Port Angeles, at the start of the Hurricane Ridge road. Hoh Rain Forest Visitor Center is located in the Hoh River Valley, with vehicular access from Highway 101 at the western edge of the park. Information stations are located (clockwise) at Ozette, in the northwestern corner, Lake Crescent, Sol Duc, Storm King, Elwha, Hurricane Ridge, Staircase, Quinault, Kalaloch, and Mora. Each of these stations offers information services, exhibits, audiovisual programs, and a sales outlet; bird field guides and a checklist are available.

Interpretive activities vary at each of these locations; many include nature walks and evening programs. Further details on locations and times are available at all the stations and on campground bulletin boards. In addition, the Olympic Park Institute offers seminars for adults and children on a variety of cultural and natural history topics. Further information can be obtained from the Olympic Park Institute, 111 Barnes Point Road, Port Angeles, WA 98362; (360) 928-3720.

Additional park information can be obtained from the superintendent, Olympic National Park, 600 East Park Avenue, Port Angeles, WA 98362; (360) 452-4501.

BIRD LIFE

Common Ravens monitored our progress along the Hurricane Hill Trail, soaring by now and then as if to check on our whereabouts. Their all-black plumage, large size, heavy bills, and wedge-shaped tails helped to distinguish these Corvids (members of the crow family) from the smaller Northwestern Crows, common at lower elevations.

We sat on a rocky outcrop near the summit admiring the grand scenery so abundantly obvious in all directions. Klahhane Ridge ran off to the northeast, hiding most of the town of Port Angeles. The deeply forested Elwha Valley was visible to the south. Mount Olympus, the park's highest peak (7,965 ft.), lay hidden by the Bailey Range to the southwest. The Queets, Hoh, Bogaciel, and Boracheil Rivers drain the Baileys to the west.

Dark-eyed Juncos searched the meadow nearby. A flock of Red Crossbills passed over, calling loud "jip-jip" notes. A pair of American Pipits flew by. I wondered where these birds were going; they nest in wet highland meadows and perform display flights similar to those of Horned Larks. I also searched the edges of the snowfields for Gray-crowned Rosy-Finches, but found none. Bruce, the park's wildlife biologist, told me that he only occasionally finds this species in summer but had seen it on Hurricane Hill in the past. Then, just below our perch, a robin began to gather nesting material, just like the one in the forest at the parking area almost 1,000 feet below. We watched it fly away with a full load of materials. I wondered aloud if this species, which occurs at all elevations, might be the park's most common bird. Bruce said the juncos were probably more numerous.

Hurricane Ridge and the various trails were crowded with visitors when we returned to our vehicle and passed through that area. Northern Flickers, Common Ravens, Violet-green Swallows, Steller's Jays, American Robins, Dark-eyed Juncos, and Pine Siskins were all evident from the parking area. And along Obstruction Point Road (still closed by snow after 4 miles), we encountered an Olive-sided Flycatcher and several Winter Wrens, varied thrushes, and Townsend's Warblers by walking a section of the roadway that passed through a mixed conifer forest. The Olive-sided Flycatcher, singing loud "pee-de" or "me'n-you" songs (often translated as "hic-three-beers"), sat at the very top of a tall snag. Its upright stance and white patches on its flanks were obvious even at a distance.

The **Townsend's Warblers** were singing wonderful, somewhat buzzy "jeer jeer jeer je-da" songs, which rise gradually at first and then ascend sharply with the "je-da." Other interpretations include "weezy weezy weezy weezy seet" or "tsooka tsooka tsooka tsook tsee! tsee!" And Hoffmann transcribed its song as "a hoarse 'swee swee swee zee.'" All of these descriptions fit. It took us several minutes of searching the high foliage of Douglas-firs before we found the source. A beautiful male, and Olympic's brightest, most colorful warbler: black throat, cap, and cheeks, bordered with bold yellow markings, and yellow flanks, heavily streaked with black; females are duller versions with a yellowish throat.

Townsend's Warblers usually are found in the high foliage during the breeding season, although they commonly occur in lower areas, often in the company of other songbirds, at other times of the year. Naturalist Fred Sharpe, in an unpublished manuscript titled "Olympic Peninsula Birds: The Songbirds," points out that after nesting this species "wanders widely and can be found in nearly any coniferous habitat from sea level to treeline." He also claims that "Townsend's is one of our hardiest warblers, and usually lingers in fair numbers in the lowlands until the end of November. On mild years individuals or small groups will spend the entire winter, particularly in the south coast region. . . . In late March and the first pulse of singing, brightly plumaged males start moving north along the outer coast. Heavy reinforcements start passing through by late April and most birds are on their mid-montane breeding areas by mid-May."

Olympic's mixed conifer forests blend with the riparian habitat along the numerous valley bottoms. One of the best examples of this community can be found in the lower Elwha Valley. Common songbirds found at Elwha and Altaire campgrounds in summer include the Pacific-slope Flycatcher, Steller's Jay, Chestnut-backed Chickadee, Golden-crowned Kinglet, Winter Wren, American Robin, Yellow-rumped Warbler, Dark-eyed Junco, and Song Sparrow. Fewer numbers of the Violet-green Swallow, Varied Thrush, Cedar Waxwing, Orange-crowned and Black-throated Gray Warblers, and Evening Grosbeak can usually be seen as well.

Of all these songbirds, the **Steller's Jay** may be the most obvious. Few birds are so daring and obnoxious as this high-crested, deep blue jay. Members of this species are especially bold at campsites, where they solicit handouts despite the potential dangers this habit portends; such nonnative foods not only lack the natural elements they require, but the resultant increase in jay populations can also pose a threat from disease.

Sharpe points out that this "incandescent blue flash" inhabits all the habitats of the Olympics. He quotes a general description of the Steller's Jay from Dawson (1909):

> Mischief and the blue jay are synonymous. Alert, saucy, inquisitive, and provoking, yet always interesting, this handsome brigand keeps his human critics in perpetual see-saw between wrath and admiration. As a sprightly piece of nature, the Steller's Jay is an unqualified success. As the hero-subject of a guessing contest he is without peer, for one never knows what he is doing until he has done it, and none may predict what he will do next."

BELOW THE HIGHLANDS

The Elwha River supports a number of breeding waterbirds: the Harlequin Duck, Common Merganser, Spotted Sandpiper, Belted Kingfisher, and American Dipper. The **Harlequin Duck** is the most colorful, although it is resident on the Elwha only from late winter to mid-summer. The rather gaudy males are present only during the first half of that period. Olympic's Harlequin males first move up the streams while the banks are still snowbound, locating nesting territories along the swift-flowing waterways. Nests, constructed of dry twigs and leaves and lined with white down, are placed on the ground among protective boulders, under bushes, or in a tree cavity. However, once the eleven or so eggs are laid, the drake joins other bachelors at sea, leaving the remainder of the family chores to the hen. She must complete incubation and care of the fledglings, including educating the youngsters to the many wiles of life, until they return to the sea for the winter months. During June and July, the drab-colored hens and youngsters can often be found feeding in the river shallows or resting on rounded boulders or logs.

Except for the double white spots on their cheeks and a noticeably round spot behind the eyes, they are an overall grayish brown color.

Male Harlequins, however, are one of our most colorful ducks. They possess an overall slate blue body with chestnut stripes on the head and sides and white markings scattered here and there over the body, including a large crescent patch on the face and stripes and spots on the neck, and a collar that almost fully encircles the head. The name "harlequin" was derived from a comical character of the Italian theater who wore a mask and multicolored tights. And its scientific name—*Histrionicus histrionicus*—comes from the Latin *histro*, or actor, suggesting the make-up worn by an actor, according to Edward Gruson's *Words for Birds*.

Olympic National Park is one of the best places in the world to observe this colorful duck, and the Elwha Valley, because of its accessibility, is the best place in the park to look. More than 360 breeding pairs occur in Olympic and about 1,300 overwinter along the Washington coastline, according to Greg Schirato of the Washington Department of Fish and Wildlife, who censuses the state population.

Common Mergansers were also reasonably common, and, as with Harlequins, only the hens and youngsters were present; the drakes had already retired to bachelor flocks. Spotted Sandpipers called "wheet wheet wheet" as they flew stiff-winged from one river boulder to the other. Several Belted Kingfishers flew along the river, giving loud rattle-calls for no apparent reason. And above the campgrounds, I located an American Dipper flying upriver with a billful of insects that it had found on the stream bottom, no doubt en route to a nest and youngsters.

Several other bird species were encountered along the 2.5-mile-long roadway to where it climbs into the forested highlands. I had stopped along the road to bird an open patch of alders when I discovered an open area below the road; a landslide had taken out the large conifers, and numerous alders and shrubs were invading. A family of Dark-eyed Juncos, at least five youngsters and their parents, were foraging on and about the shrubs in good view. The parents were trying to ignore their solicitous offspring. When a pair of Evening Grosbeaks came close, the male courting his lady with much wing-drooping and quivering, two of the young juncos began to beg food from the much larger male grosbeak. Although he repeatedly turned his back on the incessant juncos, turning away each time one managed to get in his face, he was unable to lose the youngsters. Between attempts to court and trying to ignore the juncos, he finally gave up and flew up to an adjacent snag. He then gave them a series of strident "clee-ir" calls.

Also found in that little clearing were a female Rufous Hummingbird, which buzzed the juncos a number of times; several Chestnut-backed Chickadees; the omnipresent American Robins; Winter Wrens singing from the adjacent undergrowth; singing Orange-crowned, Yellow-rumped, and Townsend's Warblers; a pair of courting Western Tanagers; a male Black-headed Grosbeak; and a pair of Song Sparrows. And just beyond, along the roadway, we watched a Blue Grouse feeding on buds.

One of the most accessible places to find lowland birds is about the grounds of the Pioneer Memorial Museum and Visitor Center. Most numerous of these are the California Quail, Rufous Hummingbird, Northern Flicker, Pacific-slope Flycatcher, Violet-green Swallow, Steller's Jay, American Robin, Swainson's Thrush, Warbling Vireo, Dark-eyed Junco, Song Sparrow, and Pine Siskin. Watch also for the Red-tailed Hawk, Band-tailed Pigeon, Common Nighthawk, Vaux's Swift, Bushtit, Winter Wren, Cedar Waxwing, Cassin's Vireo, Orange-crowned and Wilson's Warblers, Western Tanager, Spotted Towhee, House Finch, and American Goldfinch.

THE PACIFIC SLOPE

Olympic's Hoh Rain Forest is, without doubt, one of North America's most remarkable communities. This lush environment supports the greatest abundance of plant life found north of the tropics. In *Olympic Ecosystems of the Peninsula*, Michael Smithson reports that the northwestern rain forest has "more biological material than any ecosystem in the world—as much as 1 million pounds per acre." He also claims that "over 130 species of epiphytes," plants that grow on other plants, have been found in the Hoh Rain Forest. And according to Moorhead's book, *The Forest Elk*, there is no better place to find Roosevelt elk than in the Hoh Valley.

Conversely, the Hoh Rain Forest does not support a large avian population. Most of the birds that do occur there reside primarily in the canopy or in openings. During one morning of birding this area, I recorded fewer than two dozen species. Most numerous were the Pacific-slope Flycatcher, Chestnut-backed Chickadee, Golden-crowned Kinglet, Winter Wren, and Black-throated Gray Warbler. Less abundant rain forest species included the Hairy and Pileated Woodpeckers, Northern Flicker, Northwestern Crow, Steller's Jay, Red-breasted Nuthatch, Brown Creeper, Hermit and Varied Thrushes, and American Robin. Birds found only in clearings of deciduous vegetation or soaring overhead added the Osprey, Bald Eagle, Vaux's Swift, Northern Rough-winged Swallow, Common Raven, Cedar Waxwing, the nonnative European Starling, Warbling Vireo, and Song Sparrow.

None represents this deep, dark, dank forest better than the tiny **Winter Wren.** Yet it somehow seems inappropriate to these great cathedral-like forests; it is only a 4-inch, reddish brown bird with a stubby tail, which is far more often heard than seen. When walking either the Hall of Mosses or Spruce nature trails, there was never a time when one or more songs of this happy songster were not in evidence. Its song is a rapid series of high, tinkling trills and warbles. I was reminded of Arthur Cleveland Bent's description from his Life History series. He writes that its "variety is entrancing; the full rich song fairly bursts upon the area with a tinge of nature's wildness."

Black-throated Gray Warblers were surprisingly abundant in both the conifers and deciduous vegetation in and around the rain forest. They were usually detected first by their buzzy songs that oftentimes sounded very much like a Townsend's Warbler song. Sharpe states that, although they have "a standard 'wee-ze wee-zy wee-zy weet' call which is routinely given," they also sing "a series of 'wee-zy' phrases rising directly up the scale, with emphasis on the last note." Bent, on hearing the bird in Washington, described its song as "swee, swee, ker-swee, sick," or "swee, swee, swee, per-swee-ee, sic." Hoffmann points out that "even an expert gives a very cautious verdict [on song identification] in western Washington, where Hermit, Townsend's and Black-throated Gray Warblers are all found breeding."

Black-throated Grays are not so nervous as most other warblers but go about their business in a methodical and deliberate manner, very much like vireos. Therefore, close observations of this lovely warbler are often possible. It is a black-and-white bird with a black cap, cheeks, and throat, bold white stripes above and below the eyes, and a tiny yellow spot in front of each eye.

Throughout our stay in the Hoh Rain Forest I searched the moss-draped trees for sightings of the elusive and controversial Spotted Owl or Marbled Murrelet, but to no avail. These two birds are considered "threatened" and, because they require old-growth forests for nesting, have caused a great deal of controversy in the timber-oriented Northwest. Murrelets are robin-sized seabirds that nest on high mossy limbs of mature conifers up to 55 miles from the sea; only ten have been found in Washington state.

A pair of **Spotted Owls** does reside in the Hoh Rain Forest. I met owl biologist Dave Clarkson there; he was monitoring their whereabouts. Clarkson told me that in 1993, the second year of a three-year inventory and monitoring project, twenty pairs and twenty-three single owls were found within the 20,111 acres surveyed. From all evidence gathered to date, it appears that Olympic Park has the greatest number of Spotted Owls of any national park (estimated at about 150 pairs), and that the Hoh Rain Forest is one of the choicest sites for this nighttime raptor.

The Northern Spotted Owl has a distinct preference for old-growth forests, with large, multilayered, overstory trees, high canopy closure, and numerous cavities. Favorite prey species in Olympic include flying squirrels, wood rats, and white-footed mice. It is most often detected at night by its rather distinct calls: "series of 3–4 hoots (males, deep, mellow; female, high penetrating)—who-who . . . WHOOo; whup . . . who-who . . . WHOOs. or who . . . hu WHO . . . whoOOo—leisurely pace last note longer, accented," according to Julio de la Torre, in *Owls: Their Life and Behavior*. In extremely rare cases when one is found during the daytime, it can be very confiding, allowing a close approach, and therefore easy to identify. It is our only large (16–19 in. or 41–48 cm) noneared owl with all-dark eyes. It is brown overall with many whitish spots and a barred belly.

Just below the Hoh Rain Forest Visitor Center, along the entrance road, is a pond that provides nesting habitat for Mallards and Hooded Mergansers. My wife, Betty, videotaped seven Mallard hens there, each

Fig. 5. Spotted Owl

with five to a dozen youngsters in tow, and a lone Hooded Merganser hen with three large youngsters. Overhead, thirty or more Northern Rough-winged Swallows and a dozen Vaux's Swifts were plying the air in their never-ending search for insects.

Another morning, Bruce and I visited the Ozette area, birding the campground and trailhead and then hiking to Cape Alava. The most abundant birds found in the riparian setting at the trailhead included Vaux's Swifts and Tree, Rough-winged, and Barn Swallows flying overhead; Steller's Jays, Northwestern Crows, Chestnut-backed Chickadees, and Golden-crowned Kinglets about the camping area; American Robins about the lawns; Orange-crowned Warblers singing from the adjacent willows and alders; and Song Sparrows at the edges of the lake and stream. Other species recorded there included several Rufous Hummingbirds, one male actively displaying over a twinberry shrub; a family of Gray Jays that contained adults and three very dark, almost blackish youngsters; a singing Warbling Vireo and Yellow and Wilson's Warblers; and a pair of Common Yellowthroats calling loud "witchity witchity witchity" from the streamside.

"Accipiter!" Bruce suddenly called. I turned toward the lake just as a Sharp-shinned Hawk, probably a female based on its large size, sailed into the forest a few hundred feet down the lakeshore. It seemed to trigger other birds into flight, as a flock of twelve Red Crossbills arose from the forest, calling loud "jib jib" notes. Five Band-tailed Pigeons also flew out of the forest and made a high, wide circle, before gliding back into hiding. And high overhead a loon, probably a Common Loon, flew in a double circle pattern before returning perhaps to a nest hidden along the shoreline.

As we crossed the bridge at Ozette, the start of our 3-mile hike to Cape Alava, we encountered an additional assortment of forest birds: Hairy and Pileated Woodpeckers, Pacific-slope Flycatcher, Red-breasted Nuthatch, Swainson's and Varied Thrushes, Winter Wren, Dark-eyed Junco, and Purple Finch. One of the **Pacific-slope Flycatchers** sang its short, snappy "pee-ist," from an open alder near the bridge; it provided a wonderful view. A tiny, rather secretive bird with a yellowish breast, yellow-brown upperparts, a distinct white eye ring, and whitish wing bars, it jerked its tail upward with every song and movement.

This little *Empidonax* flycatcher is found in almost every habitat below the tree line, in coniferous and deciduous areas alike. Sharpe refers to it as the "ultimate 'supertramp' flycatcher of the Olympic Peninsula." He also reports that it uses a wide range of nesting sites, including "trees, on overgrown stumps, in cavities of a snag, among the roots of fallen trees, or on the electrical box of a summer cabin."

The Ozette coastline is typical of the forest-lined coastal areas of the Olympic Peninsula, replete with log-strewn beaches, boulders, and low rocky nearshore islands, including larger Ozette Island. This is where we found male Harlequin Ducks; loners and small groups were scattered among the surf or sitting on rocks. Eighteen males were crowded on three rocks close enough to shore for us to see their multicolored markings well through binoculars. Three Great Blue Herons were perched nearby. Other surfbirds included Pacific and Common Loons; Double-crested, Brandt's, and Pelagic Cormorants; Surf Scoter, Common Merganser, Caspian Tern, and Pigeon Guillemot.

The rocky outcrops and seastacks farther out contained a somewhat different assemblage of birds, including all three cormorants as well as the Glaucous-winged Gull, Black Oystercatcher, Common Murre, and Tufted Puffin. Of all these seabirds, the most numerous were the **Glaucous-winged Gulls**; a couple hundred or more appeared to be nesting on Ozette Island. These large, all-white gulls, except for a gray mantle and yellow bill, were the most dominant birds by far. A few individuals had darker mantles, suggesting hybridization with Western Gulls, a fairly common occurrence along the Washington coast. Gulls are notorious for interbreeding with other so-called species, making some gull identification extremely difficult. Paul Ehrlich and colleagues comment on this issue in *The Birder's Handbook*: "When you are in the field trying to sort out which of these gulls you have in your binoculars, take heart. The gulls themselves also have problems telling who is who."

Almost as common as gulls, **Northwestern Crows** were scattered along the shoreline and in the adjacent forest. Hoffmann writes that "the small black forms and hoarse cries of this crow are characteristic features of the [Washington coastal] landscape." These American Crow look-alikes possess many of the same characteristics, differing only in their hoarse "caar" calls, faster wingbeats, and apparent affinity for seafood. They are famous for breaking mollusks by dropping them onto rocks from above. Ehrlich points out that "yearlings take smaller clams and are less efficient at opening them." He also claims that "adults with helper (only one per territory) produce more young than adults without helper."

A flock of about forty-five crows suddenly passed overhead, between our perch on the beach and the forest, and continued up the beach for a few hundred yards. Their flock seemed more compact than the usual trailing flock of American Crows, turning and circling en masse. We watched them flock to the beach next, where they seemed only to land and regroup before flying up again toward the forest. Soon they were back over the beach and heading out to Ozette Island. There they circled a couple times before heading south over the kelp-covered rocks. Then they returned to Ozette Island, where they landed in the forest. We decided that we had been watching a multifamily training exercise in togetherness.

Throughout our stay at Cape Alava, an adult **Bald Eagle** was perched on one of the low rock islands south of Ozette Island. Although it flew off several times, each time it returned to the same perch. Other Bald Eagles, two adults and at least three juvenile birds, were also observed over the forest or surf. The adults were gorgeous creatures, majestic with their all-white heads and tails. Bruce told me that the five active eagle nests on Cape Alava had produced seven fledglings in 1994, and that eagle biologists Anita McMillan and Erran Seaman had located a total of thirty-six Bald Eagle territories along the 57-mile-long Olympic park coastline. Bruce also told me that Olympic's Bald Eagles prey on the abundant waterfowl along the coast: "I have actually watched an eagle capture a scaup in the surf and carry it to a nest."

Many of the same birds, as well as a few additional seabirds, also occur along the coast to the south near Mora. The larger seastacks off Rialto Beach provide habitat for nesting storm-petrels, cormorants, gulls, and alcidae. Terry Wahl and colleagues summarized seabird nesting status in Washington at a 1990 Pacific Seabird Group symposium. They listed thirteen species and estimated their populations for Washington: 3,900 Fork-tailed Storm-Petrels, 36,000 Leach's Storm-Petrels, 560 Brandt's Cormorants, 2,150 Double-crested Cormorants, 2,640 Pelagic Cormorants, 16,500 Glaucous-winged Gulls, 7,902 Caspian Terns, 3,000 Common Murres, 350 Pigeon Guillemots, 2,400 Marbled Murrelets, 88,000 Cassin's Auklets, and 23,700 Tufted Puffins.

Although some of these birds occasionally come close enough to shore to identify, good observations usually require a trip offshore. Such pelagic birding trips can provide opportunities to view an additional twenty or more seabirds. Wahl and colleagues point out that seabirds "appear to concentrate at fronts, outflows, areas of upwelling, and seasonal prey concentrations."

We sat on logs at the beach for a considerable time watching the comings and goings of the hundreds of gulls and fewer terns along the surf, while the sounds of various songbirds echoed from the forest. It seemed odd to find American Robins running here and there along the wet sand, just as they do over the lawns at home, searching for worms and other invertebrates.

During the fall and spring months, the beach scenes are very different. At those times, millions of migrant shorebirds can be found along the beaches and mudflats. And the offshore waters and larger lakes can be filled with waterbirds. Eugene Wilhelm, in an out-of-print booklet titled *Common Birds of Olympic National Park*, provides additional information on migratory periods:

> Olympic National Park is strategically located on the Pacific Flyway, a main avenue of flight in bird migration. The passage of birds along the Olympic Ocean Strip is equally heavy spring and fall, and follows the same pattern. . . . Migratory flights of waterfowl (geese and ducks), and seabirds (gulls, terns, cormorants, and alcids) usually reach their peaks within the following dates: spring, March 15–May 15; fall, September 30–November 1. Shorebirds are found in greatest numbers in the late summer. Dates: spring, April 30–June 1; summer, August 1–September 15.

The annual Christmas Bird Counts provide the best perspective on the winter populations. In December 1997, a total of 53,832 individual birds of 136 species were tallied on the Sequim-Dungeness, Washington Count. The dozen most numerous water and terrestrial birds, in descending order of abundance, included (waterbirds) American Wigeon, Mallard, Dunlin, Northern Pintail, Bufflehead, Red-breasted Merganser, Surf Scoter, Oldsquaw, Green-winged Teal, Common Goldeneye, Sanderling, and Mew Gull; and (terrestrial birds) European Starling, Pine Siskin, Red-winged Blackbird, Brewer's Blackbird, American Robin, Dark-eyed Junco, Golden-crowned Kinglet, House Finch, Bushtit, Spotted Towhee, Chestnut-backed Chickadee, and Ruby-crowned Kinglet.

In summary, the park's checklist of birds includes 355 species, of which 145 are known to nest. Of those 145 breeding birds, 37 are waterbirds (grebe, seabirds, waders, waterfowl, rails, shorebirds, gulls, and terns), 19 are hawks and owls, and 9 are warblers.

BIRDS OF SPECIAL INTEREST

Harlequin Duck. The multicolored males frequent swift-flowing streams from late March to early June, then spend their summers at sea, leaving the duller hens to raise the chicks.

Bald Eagle. This majestic white-headed raptor is most common along the coast and inland along the major rivers during salmon runs.

Glaucous-winged Gull. The park's most abundant large, resident gull. Breeding adults have white heads and underparts, gray mantles, and yellow bills with a reddish spot on the lower mandible.

Spotted Owl. About 150 nesting pairs of these endangered birds reside in the park's old-growth forests; it is a large, noneared owl with all-brown eyes.

Pacific-slope Flycatcher. This is the little, yellowish flycatcher of the conifers and deciduous thickets that sings a sharp "pee-ist" note over and over.

Horned Lark. Watch for this "horned" bird in high meadows; it performs a high-flying display while singing its tinkling song.

Steller's Jay. This is the all-dark-blue bird with a high, blackish crest and aggressive behavior; it has a harsh "shaack" call.

Northwestern Crow. The park's only crow, it is especially common along the coast but can be expected almost anywhere below the tree line.

Winter Wren. This tiny, reddish brown bird with a stubby tail resides in the forest undergrowth and sings a continuous song, like tinkling bells.

Black-throated Gray Warbler. Most common in the rain forest, it sports a black-and-white head pattern with a tiny yellow spot in front of each eye.

Townsend's Warbler. One of the park's most colorful birds, it has a black-and-yellow head pattern.

North Cascades National Park and Ross Lake
and Lake Chelan National Recreation Areas,
Washington

The granite spires of the Picket Range provide a marvelous backdrop to the montane forest visible from the viewing terrace behind the North Cascades Visitor Center. The highest point visible is Mount Terror, at 8,151 feet, although Mount Degenhardt (8,030 ft.), to the right of center, is closer and appears to be higher. Mount Challenger (8,248 ft.) and Challenger Glacier are hidden from view. And Mount Baker, the highest peak (10,778 ft.) in the North Cascades, is far to the northwest.

One morning in late June, I stood at the visitor center terrace, admiring the distant peaks, many with glistening snowfields. A chickaree, or Douglas squirrel, called from the forest to my left. An American Robin sang a cheerful song from a tall fir tree beyond. A Pacific-slope Flycatcher called sharp "pee-ist" notes nearby. Behind me, Steller's Jays were scolding some unfortunate creature that they had happened upon. And then, directly below my perch, a Dark-eyed Junco appeared out of the undergrowth. I watched it work its way upward, from shrubbery to treetop, where it suddenly began to sing its melodic trill. Its all-black hood, brownish back and sides, and black tail with white outer tail feathers, were all obvious in the morning light. With binoculars I could also see its tiny, pinkish bill.

A little yellow-and-black bird flew into the foliage of a conifer some distance beyond. It immediately began to sing a rather weezy but lovely song, with three rising notes followed by two sharp descending ones: "bzeee bzeee bzeee, tsee-see." It was a male **Townsend's Warbler,** one of the park's most gorgeous creatures, and one that nests in all of the park's forested zones. I zeroed in with my binoculars just as it put its head back and sang, "bzeee bzeee, tsee-see," a slightly different rendition of its basic song. Its coal black throat, cheeks, and cap contrasted with the bold, canary yellow lines above and below the cheeks. I watched as it foraged among the tree boughs for insects, constantly moving from one to another. And every now and then it would stop, put its head back and sing: "bzeee bzeee bzeee, tsee-see."

Two black objects passed through my viewing field some distance beyond. Shifting my attention away from the warbler, I discovered a pair of **Black Swifts** in a wild chase, turning, tumbling, diving, and then ascending straight up with incredible speed. Their very long, stiff, narrow wings, beating at different speeds in turns, gave their flights a twinkling effect. The smaller and slower White-throated Swift of the American Southwest has been clocked at 110 miles per hour. At one point, when the two Black Swifts came a little closer, the undersides of their wings seemed silvery, and I could also see their slightly forked tails. Their call was a high-pitched "tic tic tic."

As I watched these incredible birds in the magnificent setting of the North Cascades, I realized that the Black Swift, normally hard to find elsewhere in its range, is this park's best representative. At North Cascades, Black Swifts can be seen from the high glaciated canyons that symbolize the North Cascades to the crystal-clear airshed over the deep blue lakes and green forests.

THE PARK ENVIRONMENT

North Cascades National Park and the two interrelated national recreation areas encompass an area of about 648,000 acres, of which 93 percent is officially designated as the Stephen Mather Wilderness Area. The nonwilderness portion includes a central corridor that encompasses Highway 20 and three dammed lakes within Ross Lake National Recreation Area; a portion is administered by Seattle City Light. The national park complex is surrounded by additional outstanding scenery and wildlands administered by the U.S. Forest Service and British Columbian Provincial Parks.

Fig. 6. Black Swifts

The national park brochure claims that "North Cascades National Park contains some of America's most breathtakingly beautiful scenery—high jagged peaks, ridges, slopes, and countless cascading waterfalls." It further states that the area "encompasses some 318 glaciers, more than half of all glaciers in the contiguous United States." And Henry Custer, a nineteenth-century topographer, wrote: "Whoever wishes to see nature in all its primitive glory and grandeur, in its almost ferocious wildness, must go and visit these mountain regions." I most certainly agree.

Area vegetation was studied by James Agee and Jane Kertis, who identified eighteen plant cover types that can be lumped into five rather distinct communities: riparian vegetation, dominated by hardwoods such as bigleaf maple, black cottonwood, and red alder, with an understory of thimbleberry, blackberry, and bracken fern, occurs in the river valley bottoms and lower slopes; red alders and willows dominate the higher streamsides. A ponderosa pine community occurs on the drier eastern slopes. Montane forests of conifers or conifers mixed with hardwoods occur above the valley bottoms and extend up to the subalpine zone; Douglas-fir is most abundant, although lodgepole pine, western red cedar, western hemlock, subalpine fir, and Engelmann spruce may all be present in varying numbers. The subalpine community is dominated by subalpine fir with fewer numbers of mountain hemlock, silver fir, and Alaska-cedar. Finally, the alpine community consists of bare rock and scattered shrubs, such as heathers, and wildflowers.

The National Park Service operates the North Cascades Visitor Center at Newhalem and information stations at Marblemount, Stehekin, Glacier, and Chelan. Each of these has an information desk, exhibits, and a sales outlet; bird field guides and a checklist are available. The visitor center also has an auditorium for orientation and other programs. Nature walks and evening talks are provided by park interpreters during the summer at Colonial Creek and Newhalem Creek campgrounds, and at Hozomeen and Stehekin. Schedules are posted at the campgrounds and available at the visitor center.

In addition, the North Cascades Institute, a nonprofit field school, offers a wide assortment of classes, including seminars on Okanagan and San Juan Island birds, fall raptor migration, and birds of winter, as well as spring field ornithology classes. Details are available from the North Cascades Institute, 2105 Highway 20, Sedro Woolley, WA 98284.

Additional information on the national park complex is available from the superintendent, North Cascades National Park, 2105 Highway 20, Sedro Woolley, WA 98284; (360) 856-5700.

BIRD LIFE

Black Swifts are featured in a visitor center exhibit that includes a replica of a nest with one tiny baby. The text with the exhibit reads: "Black Swifts fly up to 600 miles (966 km) a day in search of insects to feed their one chick. While the parents are gone the chick waits, sometimes 2 or 3 days, in torpor, a state of prolonged sleep. Its breathing and heart rate drop dramatically to conserve energy, down to one breath and four to eight heart beats a minute."

One reason this bird is so numerous at North Cascades is its practice of nesting near or behind waterfalls, and few parks can compete with North Cascades for the number of spectacular waterfalls. Colonial nesters, breeding Black Swifts gather moss, algae, ferns, and other nest materials in flight, which they glue together with their saliva. They use conifer needles and fine rootlets to line their nests.

Swifts never perch on vegetation or wires or sit on the ground. Their very tiny, weak feet are adapted only for clinging onto cliffs. Except when incubating, their entire existence is spent on the wing. They feed, perhaps even sleep, and mate in flight. Copulating swifts, clinging together with wings out to slow their descent, may fall, spinning round and round for hundreds of feet, resuming flight only a few feet above the ground. And John Terres, in *The Audubon Society Encyclopedia of North American Birds*, reports that in summer the Black Swift will "flock at edge of clouds, follows cloud for hundreds of miles to feed in warm insect-laden air mass; from this habit often called cloud swift; on clear days may ascend until invisible at height of several thousand feet."

I also sampled the montane forest bird life during a morning walk along the Thunder Creek Trail. Starting at the Colonial Creek parking area, at the edge of the campground, I discovered a dozen or more

Black Swifts soaring overhead. Apparently they had located a mass of insects and were feeding 300 to 800 feet above. All the while, several American Robins and Swainson's and Varied Thrushes sang their distinct songs from the adjacent vegetation. A lone Hermit Thrush song joined the thrush choir from the conifers above the parking area. I couldn't help wondering if there is a place in Stehekin Valley, in the southeastern corner of the park complex, where the Veery also can be added to the thrush choir. Bob Kuntz, park wildlife biologist, later told me that Veeries are indeed reasonably common in the deciduous and mixed forests in the Stehekin.

I know of no other park where all five thrushes occur together. Each has a wonderful song, easily separated from the others. Robins and Swainson's Thrushes are most numerous. Robins sing a rich, cheerful caroling song that is often written as "cheerily-cheery-cheerily-cheery." The very different song of the Swainson's Thrush is an "upward-rolling series of flutelike phrases, like "wip-poor-wil-wil-eez-zee-zee," as described by Wayne Petersen in The Audubon Society Master Guide to Birding. Veeries sing a breezy song that descends: "da-vee-ur, vee-ur, veer, veer."

The widespread **Varied Thrush** sings a wonderfully unique song that Roger Tory Peterson, in A Field Guide to Western Birds, describes as a "long, eerie, quavering whistled note, followed, after a pause, by one on a lower or higher pitch." I have imitated this song by whistling it at the highest possible pitch.

The Varied Thrush is unquestionably one of the park's most charismatic songbirds. Its eerie songs often echo through the forest from hidden places, giving the listener the feeling either of mystery or enchantment, depending upon one's perspective. Slightly larger and, at first sighting, seemingly similar to the American Robin, it possesses very different plumage: orange throat and eye stripe against a black cap and cheeks, and blackish back and wings with bold orange wing bars. Its reddish orange chest is crossed by a bold black band. But like the robin, during the breeding season it sings from the first light of day until dark.

The Hermit Thrush is a bird of the conifers, just as the Swainson's Thrush seems to prefer deciduous vegetation; both may be found in mixed forests. High-country campers and hikers are most likely to experience the Hermit's incredible music. Terres provides us with the best description: it "opens with clear flutelike note, followed by ethereal, bell-like tones, ascending and descending in no fixed order, rising until reach dizzying vocal heights and notes fade away in silvery tinkle."

A much larger, white-and-black bird suddenly appeared high overhead, flying up Thunder Arm. Through binoculars, I was quickly able to identify it as an Osprey that was carrying a fish held firmly with its talons against its belly. The bird's long wings, white underparts, black corners at the bent wrists, and dark back were obvious. Undoubtedly it had a nest somewhere up the drainage and was taking food home to a waiting family.

Other birds evident from the parking area and campground included the Pacific-slope Flycatcher, Gray and Steller's Jays, American Crow, Winter Wren, Cedar Waxwing, Warbling Vireo, Townsend's Warbler, and Song Sparrow. The most obvious of these was the high-crested **Steller's Jay,** a blue jay with a blackish head and obnoxious personality. Although these birds occur throughout the area, they are most numerous in the vicinity of the campgrounds, where they become true camp robbers, begging for handouts or stealing snacks right off the table. This bird, a member of the crow family (Corvidae), will gather food so long as it is available. Most of the snacks are cached on trees for later use.

The dominant bird of adjacent Diablo Lake, and even in the still waters of Thunder Creek, was the **Canada Goose.** These large water birds were surprisingly abundant. At one point, Betty counted more than thirty-five individuals of which about twenty were fledglings. We marveled at the attention both parents gave their youngsters, keeping careful watch for any threat and herding them away from any potential danger. Adults were distinctly marked with a typical black-stocking head and neck and a white cheek patch. Several Barrow's Goldeneyes and a lone Common Merganser, with three chicks, were also found in the upper bay.

As soon as I entered the forest, just beyond the amphitheater, Winter Wren songs dominated the shadowy undergrowth. There was seldom a time along the Thunder Creek Trail that their rapid, tinkling songs were not apparent. On the few occasions when I got a view of one of these tiny songsters, it was barely 4 inches (10 cm) in length, with a stubby tail, all reddish brown, and never still. Watching their constant movement among the downed logs and undergrowth was reminiscent of watching field mice. They poked and pried into every hole and cranny they came upon. Nests, constructed of mosses, grasses, twigs, feathers, hair, and the like, are built in rotted stumps or in roots and debris.

Other birds found along the trail included the Band-tailed Pigeon; Rufous Hummingbird; Northern Flicker; Hairy and Pileated Woodpeckers; Red-breasted Sapsucker; Hammond's and Pacific-slope Flycatchers; Golden-crowned Kinglet; Red-breasted Nuthatch; Townsend's, Yellow-rumped, and MacGillivray's Warblers; Purple Finch; and Pine Siskin. Along the trail I also encountered Roger Christopherson, a National Park Service biologist, who was surveying the Thunder Creek drainage for Spotted Owls. He had not located a single Spotted Owl but had recorded sixteen **Barred Owls.** This more aggressive species is known to displace the gentler, endangered Spotted Owl.

The long corridor that divides the park in two is dominated on the western slope by the Skagit River and lakes and the high power lines of Seattle City Light, and Ruby Creek to the east. The corridor contains extensive areas of riparian habitat along the river and adjacent ponds. This environment supports a large variety of birds; a few are rare or unknown elsewhere in the park complex. The Goodell Creek/County Line Ponds area, just below Newhalem, contains the best representatives.

Birding this area one morning in late June, I recorded more than three dozen species. Most abundant were the American Robin, Swainson's Thrush, Song Sparrow, Steller's Jay, Warbling Vireo, Varied Thrush, MacGillivray's Warbler, Black-capped Chickadee, and Pacific-slope Flycatcher, more or less in that order. Several of these already were feeding fledglings. One aggressive Song Sparrow charged out of the underbrush when I approached, calling sharp "tick" notes. It dropped to the ground near my feet, and ran around in circles dragging its wings, attempting to draw me away from its nest or fledglings.

The low, shrubby vegetation contained a lone Blue Grouse, which

Betty videotaped on a low snag; numerous Rufous Hummingbirds, including bright Rufous males that flew around me with loud buzzing sounds; two Willow Flycatchers, which sang energetic "fitz-bew" songs; MacGillivray's Warblers; Common Yellowthroats singing "witchity witchity witchity;" Spotted Towhees, with their distinct "breeee" songs; and a family of White-crowned Sparrows.

MacGillivray's Warblers were easily attracted to within a few feet by low spishing sounds. Males responded almost immediately, coming out of the undergrowth and calling loud "shik" notes. Three different females were carrying grubs or caterpillars, which hung from their bills while they waited for me to move on so that they might feed their waiting nestlings. Several males were singing from the undergrowth or low alders; their songs were loud and clear: "tswee, tswee, tswee wit wit" or "swee swee swee swee swe." I found one of the vocal males perched in the sunlight, as if it were in a spotlight on a dark stage. Lighted as it was, its dark gray hood showed a deep blue cast, highlighting its white, broken eye rings, olive-brown back, and lemon yellow underparts. A truly lovely bird!

The MacGillivray's Warbler is one of the park's Neotropical migrants; it breeds in the mountains of western North America and winters in the tropics, from central Mexico to Panama. Although much of its breeding grounds are protected in national parks, such as North Cascades, its winter homes are subject to massive clearing projects, primarily for pasturage to produce more cattle for America's fast-food industry.

The mature alder-cottonwood woodland also contained a number of interesting birds, including two species not normally found in the Pacific states: Red-eyed Vireo and American Redstart. Other, more expected species included the Downy Woodpecker, Black-capped Chickadee, and Black-headed Grosbeak. And in mixed forest areas I added the Northern Flicker, Chestnut-backed Chickadee, Winter Wren, Cedar Waxwing, Cassin's Vireo, and Yellow-rumped Warbler.

The ponds and riverway contained a very different variety of birds. Spotted Sandpipers were most numerous, calling loud "weet" as they flew from one area of shoreline to another. I also discovered a Common Merganser female with three youngsters; they were hugging the

Fig. 7. MacGillivray's Warbler

bank, trying to stay hidden as much as possible. Two sets of immature Mallards, with blue speculums edged with white, were found on other ponds.

I watched three Spotted Sandpipers for a considerable time, expecting them to reveal their nest site, but to no avail. They simply flew about with their stiff, rapid wingbeats, and did much tilting back and forth on bended knees. Three adults is a normal family group for Spotteds, as this species practices polyandry. The female selects the territory and her mates, defends them against other females, and the males tend to most of the family chores. See the chapter on Lassen for additional information about this bird's odd nesting habits.

The faster side streams also support American Dippers, which nest in protected crevices and among boulders within inches of the tumbling waters. They are our only aquatic songbirds, which find their food (mostly insect larvae) on the bottoms of fast-flowing streams. They can swim to 20 feet (6 m) below the surface while feeding. See the chapter on Mount Rainier about this bird's special adaptations for its underwater habits. Dippers, sometimes known as "water ouzels," are robin sized, plump, and all dark gray; they constantly dip or bob up and down.

The Stehekin Valley bird life, surveyed by Kuntz and Reed Glesne in June and July, is similar to that in the Skagit River valley, although seemingly there are larger numbers of birds of eastern affinity. They located forty-four species in four habitats, with eight dominant birds making up half of all detections: Hammond's Flycatcher, Swainson's Thrush, the American Robin, Red-eyed Vireo, Yellow-rumped Warbler, MacGillivray's Warbler, Western Tanager, and Dark-eyed Junco.

Kuntz and Jack Oelfke also estimated the **Harlequin Duck** nesting population on the Stehekin River at seven to eleven pairs. Male Harlequins are among our most colorful ducks. They possess a slate blue head, numerous white slashes at various places and angles, chestnut flanks, and gray-blue bodies. According to Kuntz and Oelfke, Harlequins arrive on the Stehekin River "from late March through April. Egg-laying probably occurs from late April to early June. Shortly after females begin incubation, males depart and return to coastal waters. This life history trait eliminates opportunities for renesting, if clutch failure occurs. . . . From late-August to mid-September, females and juvenile Harlequin Ducks leave the Stehekin River and return to coastal waters." See the chapter on Olympic for additional details about this lovely creature.

The glamour bird of the Skagit River is undoubtedly the **Bald Eagle,** America's official symbol. According to Kuntz, six hundred to seven hundred of these magnificent raptors winter along the Skagit, from Marblemount to Newhalem. They arrive from late November to early January during the upstream migration and spawning of chum salmon, and feed on these fish. The Nature Conservancy and the Washington Department of Fish and Wildlife maintain a 1,500-acre

Skagit River Bald Eagle Natural Area below the park near Rockport. And the local communities sponsor an annual Upper Skagit Bald Eagle Festival in January.

CASCADE PASS AREA

The true essence of the North Cascades, however, must be sampled in the backcountry, above the corridor and dammed lakes. Numerous readily accessible trails enter the park wilderness, but road access is limited. The most accessible is the Cascade River Road that begins at Marblemount along Highway 20 and terminates in 22 miles in a high subalpine valley. Cascade Pass (5,392 ft.) is 3.7 miles beyond, via a relatively easy mountain trail, which passes through montane and subalpine forest and across steep talus slopes. The scenery along the upper roadway and trail, including the high cliffs, glaciers, and abundant waterfalls, is magnificent.

The parking area at the trailhead was alive with birdsongs in late June. An Olive-sided Flycatcher sang its loud "three-beers" song from the very top of a high fir tree. Varied and Hermit Thrush songs echoed from the adjacent forest. And the alder-willow shrubbery, which dominates the interlaced streambed below the parking area, provided habitat for a number of additional birds. Fox Sparrows were singing loudest; their sweet notes, followed by trills and runs, rang out from several locations. I located one of these large, reddish-backed sparrows, sitting at the top of a nearby salmonberry shrub. Warbling Vireos sang their repetitive whistle-songs from the alders. Yellow Warblers were also actively singing their cheerful "tseet-tseet-tseet, sitta-sitta-see" songs. The punctuated "tswee, tswee, tswee, twsee wit" song of a Wilson's Warbler emanated from the alders below the forest.

I walked along the edge of the parking area, searching for whatever wildlife was present. Two miles down the road, Betty and I had found a black bear meandering across a long green slope across the canyon. Two mule deer does were stalking the parking area, apparently expecting handouts. Golden-mantled ground squirrels and chipmunks scampered about. Then, less than 3 feet away was a Blue Grouse, standing erect and alert, staring at me in defiance. Another step and it suddenly took flight, flying at me, actually hitting my arm with its wings. A second later it landed and growled low "kuk kuk kuk" sounds; two tiny yellowish chicks suddenly ran out of the grass, almost at my feet, disappearing into the alder thicket. The adult grouse slowly followed, walking in a cocky, jerky fashion. Although I had been attacked years earlier by a Ruffed Grouse in Wyoming, it was my first such encounter with a defensive Blue Grouse.

Overhead, Black and Vaux's Swifts zoomed here and there in their continuous search for insects. I wondered if the Black Swifts were nesting under any of the numerous waterfalls visible from the parking area. A pair of Vaux's Swifts, much smaller than the Blacks and with whitish underparts, spent considerable time near the top of a tall fir tree, and on one pass they disappeared from sight altogether. Apparently they were nesting in a high cavity. A number of Pine Siskins and a pair of Red Crossbills passed overhead. Then a brightly colored, male Red-breasted Sapsucker flew out of the forest and disappeared into the alders; it had a billful of insects, and I assumed it was feeding nestlings. It made two additional trips while I was watching.

Cascade Pass Trail switchbacks through montane forest for about 2 miles before coming out onto talus slopes with scattered subalpine firs and deciduous thickets. Forest birds detected here included about the same species as were found along Thunder Creek. Townsend's Warblers were especially vocal, singing an assortment of breezy songs. Several seemed to be singing a rendition that sounded most like "hea, yea, yea, sex-y."

At one point in the forest, I whistled the repetitive notes of a Northern Pygmy-Owl, like "kew, kew, kew, kew," and had an almost immediate response. I was mobbed by a number of forest birds: Pacific-slope Flycatcher, Chestnut-backed Chickadee, Golden-crowned Kinglet, Hermit and Varied Thrushes, Townsend's and Yellow-rumped Warblers, and Dark-eyed Junco. A Hairy Woodpecker also came by to check out the disturbance.

Hoary marmots were reasonably common along the trail beyond the forest. One individual perched on a rocky outcrop and called sharp whistle-notes at me. Pikas called from the open scree slopes, and when a Common Raven soared over the pass and cruised across the open slopes, eight or nine pikas took up the call.

I sat at the pass for a long time admiring the incredible scenery in every direction. The Stehekin River drainage lay before me to the southeast, and the steep slopes, still partially covered with snow, provided a contrasting perspective. A family of **Clark's Nutcrackers** was flying about the stand of firs off to my left, and I watched one adult shoving food into the mouth of a begging youngster. Then they flew off across the drainage, calling loud, drawn-out "kra-a-a" notes. I wondered if the adults had a food cache nearby. This medium-sized, gray bird, with white trailing edges on its wings and broad, white outer tail feathers, buries seeds on the ground during summer and fall. And it is able to remember the location of about one thousand seed caches from one season to another. See the chapter on Crater Lake for details about this marvelous bird.

From my high perch at Cascade Pass, I began to realize that I had been hearing high-pitched chirping sounds somewhere below me. With binoculars, I scanned the open slopes and low shrubbery. Nothing. Shifting to the right, I was suddenly staring at a little rosy bird picking at the open ground below a snowbank, 200 feet or more away. It was a **Gray-crowned Rosy-Finch,** a true alpine species that spends it summers above the tree line. It gathers together in huge flocks after nesting, however, and moves to lower, warmer elevations to the east of the Cascades for the winter months. As I zeroed in on my little rosy-finch, with a black crown and light gray nape, I discovered six others, all feeding on the ground or on low shrubs. One individual seemed to be eating buds from a patch of red heather. Then, as I watched, they suddenly took off, flying toward me and up the steep slope to my right. In passing, I detected a series of high-pitched notes, "chee-chee-chi-chi-che."

Although rosy-finches leave the high country in winter, a few other birds, such as Clark's Nutcrackers, Gray Jays, Golden-crowned Kinglets, and Red-breasted Nuthatches, normally remain on their territories, moving to lower elevations only during severe storms.

Winter birds in the park complex are surveyed annually in Christmas Bird Counts, which provide a good perspective on what species are present at that time of year. In 1997, counters tallied 669 individuals of 36 species on the North Cascades, Washington Count. Of those 36 species, the dozen most numerous birds, in descending order of abundance, included Golden-crowned Kinglet, Bufflehead, Varied Thrush, Pine Siskin, Winter Wren, Steller's Jay, Chestnut-backed Chickadee, Canada Goose, Bald Eagle, Common Merganser, American Crow, and Common Raven and American Dipper tied.

In summary, the North Cascades complex checklist of birds includes 202 species, of which 75 are known to nest. Of those 75 species, 7 are waterbirds, 8 are hawks and owls, and 8 are warblers.

BIRDS OF SPECIAL INTEREST

Canada Goose. This large waterbird, with a black-stocking neck and a head with white cheeks, is common along Diablo Lake and the head of Lake Chelan in summer.

Harlequin Duck. Watch for this duck along swift streams; males are multicolored, while females and youngsters are all brown with whitish head spots.

Bald Eagle. Common along the Skagit River in winter, they are easily identified by their huge size and the adult's all-white head and tail.

Black Swift. This is the all-black bird with long, pointed, swept-back wings that flies with a twinkling effect.

Steller's Jay. Common everywhere people gather, this crested jay is dark blue with a blackish head and has loud, harsh "shaack" calls.

Clark's Nutcracker. This medium-sized, black-and-white bird is most often seen in the high, subalpine areas; it has a loud, grating "kra-a-a" call.

Varied Thrush. This is the common songbird that sings high, eerie, quavering whistles that change pitch after a pause.

Townsend's Warbler. One of the area's most colorful forest birds; males sport a coal black throat, cap, and cheeks, bordered with bold yellow lines.

MacGillivray's Warbler. Watch for this lovely bird in moist thickets; males possess a dark gray hood, broken eye ring, and canary yellow underparts.

Dark-eyed Junco. Common at all elevations in summer, this little "snowbird" sports an all-black hood, brownish back and sides, and a black tail with white outer tail feathers.

Mount Rainier National Park, Washington

Low clouds hung over Rampart Ridge, shrouding my view of Mount Rainier. From Longmire Meadow, along the Shadows Nature Trail, I stared into the clouds, wishing for a clearing to reveal a part of the great snow-capped peak. Only gray clouds and deep green forest. Nearby, however, the meadow and the surrounding forest were alive with bird sounds. Songs and calls of American Robins, Varied Thrushes, Steller's Jays, Dark-eyed Juncos, Chestnut-backed Chickadees, Redwinged Blackbirds, and Song Sparrows were most obvious. Overhead, twittering Violet-green and Barn Swallows plied the air for insects.

Through binoculars, I identified the reddish brown figure lying at the far end of the meadow as a black-tailed deer doe. She was gazing intently at me; probably had been watching me since I first entered the meadow. Suddenly, an all-black Common Raven, evident by its large size, huge bill, and wedge-shaped tail, crossed over the far trees, flying along the edge of the clearing. A Steller's Jay discovered it first, flying up from the conifers, screeching its defiance. Two American Robins took up the chase next, and a male Red-winged Blackbird ended the encounter as the raven passed over my perch and continued on toward the National Park Inn.

A pair of Red-breasted Nuthatches was searching the lush foliage and cones of a nearby pine tree, totally ignoring my presence. Neat little birds with black caps, white eyelines, gray backs, and bright rusty underparts, they are very different from the Chestnut-backed Chickadees, which have chestnut backs and sides, white underparts, white cheeks, and bibs and caps. But both are nervous birds that are on the go continuously. Nuthatches feed almost exclusively on conifers, while chickadees frequent conifers and deciduous trees about equally.

Chestnut-backed Chickadees sing a husky, rather drawling "tsic tsic tsic tyee" song but frequently utter "check check" or "zee-zee" notes as well. They are curious and can often be called to within a few feet by low spishing sounds. Conversely, Red-breasted Nuthatches very rarely respond to odd squeaks and spishing. Usually they are detected first by their distinct nasal "nyak nyak nyak" calls, like toy horns.

The eerie, bell-like song of a **Varied Thrush** echoed across the meadow. A second later I had located this robin-sized thrush at the very top of an old snag. With binoculars I could see the bold black band across its deep-orange-colored chest, black cheeks with contrasting orange eyelines and throat, and dark gray back and wings with orange wing bars. It sang a dozen or more phrases and then was silent. Each phrase had begun on a slightly different level, but each was delivered with enthusiasm. Ralph Hoffmann, in *Birds of the Pacific States*, describes its song as a "long-drawn quavering note with something of the quality of escaping steam; after a short interval the note is repeated in a higher pitch, again in a lower, 'ee-ee-ee-ee.' The notes have a meditative quality due to their deliberation and above all a strangeness due partly to their quality and partly to the complete invincibility of the singer."

The Varied Thrush is not only one of the park's most common summer birds, but it resides almost everywhere below the tree line. Arriving in early March, it slowly follows the snow line as it retreats up the slopes. If a late snowstorm occurs, it can be found singing its unique songs from atop the snow-draped conifers. One day in mid-June, I was mesmerized by its eerie songs that echoed through the snowfall at Reflection Lakes.

THE PARK ENVIRONMENT

Few mountains possess the grandeur and scenic splendor of Rainier. It rises above the deep green forests like a great silver crown. Mount Rainier, at 14,411 feet, is the highest and most impressive of the several

Fig. 8. Varied Thrush

majestic, volcanic peaks in the Cascade Range: Adams, Baker, Hood, Jefferson, Shasta, and Lassen. Peter Farb, in *Face of North America*, writes:

> What makes Rainier so impressive is that it rises upon its immense base, covering about a hundred square miles [160 sq. km], directly out of the tidewater of Puget Sound. On its flank, like ermine robes on the shoulders of some king of mountains, lie 26 glaciers, a greater expanse of ice than on any other peak in the mainland United States. Its majestic bulk, its isolation from the rivalry of other peaks, the fact that nowhere can the summit be seen from the base—these things have made Rainier the most striking of all the mountains in the mainland United States.

A different perspective is provided by eighteenth-century poet Joseph Warton:

'Midst broken rocks, a rugged scene,
With green and grassy dales between,
Where nature seems to sit alone,
Majestic on a craggy throne.

Approximately 40 percent of Rainier's landscape lies above the tree line, a ragged border of the subalpine forest at an elevation of about 5,800 feet. Between the subalpine and ice fields are low shrub and alpine tundra communities dominated by a handful of flowering plants, such as heather, a wide variety of wildflowers, and lichens and algae. Ann Zwinger and Beatrice Willard, in *Land above the Trees*, point out that a "few mosses grow in the steam-heated crater of the summit, and lichens rim it at 14,300 feet but no vascular plants are known to exist above the 10,000-foot line."

The subalpine forest, which occasionally reaches to 6,800 feet, consists of dwarfed shrubs and trees, often in prostrate or krummholz position, including subalpine fir, Pacific silver fir, mountain hemlock, and Alaska yellow cedar. Below the subalpine zone, generally at an elevation between 3,000 and 4,900 feet, is a Pacific silver fir community; noble fir, mountain hemlock, and Alaska yellow cedar are also common, and huckleberries, bunchberry, pipsissewa, and Cascades azalea dominate the understory. Finally, below the silver fir community is a lowland forest of large, old conifers, especially Douglas-fir, western red cedar, western hemlock, and grand fir, with an understory of vine maple, devil's club, Oregon grape, and salal.

By national park standards Mount Rainier National Park is not a large park, at 235,612 acres, but 97 percent of that acreage is designated as wilderness. More than 300 miles of trails, many of spectacular nature, lead into the backcountry; they are usually snow free from mid-July through September. Primary vehicular access is limited to the park's southern, southeastern, and eastern edges, and the White River entrance road to Sunrise, at an elevation of 6,400 feet, the highest point in the park reached by road. Two secondary access roads lead from outside the park to the Carbon River and Mowich Lake areas in the park's northwestern corner.

The National Park Service operates five visitor information centers in the park in summer, at Longmire, Paradise, Ohanapecosh, Sunrise, and Carbon River. Each offers information services, an orientation program, exhibits, and a sales outlet; a bird field guide and checklist are available. The five centers also serve as a hub for a variety of interpretive activities. These range from evening campfire programs to guided walks and hikes; further details are posted at the numerous campgrounds and listed in the park newspaper, *Tahoma*, available at all entrance stations and visitor centers.

Additional park information can be obtained from the superintendent, Mount Rainier National Park, Tahoma Woods, Star Route, Ashford, WA 98304; (360) 569-2211.

BIRD LIFE

James Longmire moved to the meadow that bears his name in 1884, founding Longmire Medical Springs and a two-story hotel. The only remains of Longmire's enterprise is a small cabin, built in 1888. Nature has reclaimed the scene; even beavers have dammed a springflow, producing several ponds and waterways. Red-winged Blackbirds, Song Sparrows, and Common Yellowthroats are the most obvious benefactors. At least six pairs of redwings were present that June morning, chasing one another off their territories and singing loud, gurgling "ook-a-lee" or musical "tee-urr" notes from various posts. The males' scarlet epaulets (wing patches) contrasted with their coal black plumage; females possessed heavily streaked underparts.

On the far side of the meadow, a pair of Band-tailed Pigeons flew into the bare branches of an ancient snag. A male Yellow-rumped ("Audubon's") Warbler was fly-catching from another snag nearby. A Western Wood-Pewee sang a nasal "peeer" song beyond. And a pair of **Vaux's Swifts** suddenly joined the numerous swallows that were feeding over the meadow. The swifts' long, narrow, and stiff wings gave them a twinkling flight, very different from that of the swallows. Swifts beat their wings at different speeds in flight; in turning, one wing will beat faster than the other. Often described as flying cigars, swifts are cousins of hummingbirds, and not related to swallows at all. Vaux's Swifts nest in cavities in old-growth conifers, gathering nesting materials (twigs and conifer needles) in flight and gluing them together and to the walls of hollow trees with saliva.

I found several other birds along the three-quarter-mile loop trail. In the forest, I recorded a Hairy Woodpecker, a Northern Flicker, a Hammond's Flycatcher and several Pacific-slope Flycatchers, Golden-crowned Kinglets, Brown Creepers, Winter Wrens, and Hermit Thrushes. Among the alders at the meadow outlet, I added a pair of Red-breasted Sapsuckers and a singing Willow Flycatcher, Swainson's Thrush, and Warbling Vireo.

Finding four thrushes—the American Robin and Varied, Hermit, and Swainson's Thrushes—in such close proximity was of special interest. Robins occurred, more or less, only in the vicinity of the meadow; Varied Thrushes were present in all the habitats; Hermits were found only in the coniferous forest above the meadow; and the Swainson's Thrush occurred only among the alders. On one occasion I was able to hear both Hermit and Swainson's Thrush songs from one location—very different songs and easily identified. The Swainson's sings an upward-rolling series of flutelike phrases that Wayne Peterson, in The Audubon Society Master Guide to Birding, describes as "wippoor-wil-wil-eez-see-see."

The songs of the **Hermit Thrush** are, to my ear, among the most appealing of all. This flutelike song is described by John Terres, in The Audubon Society Encyclopedia of North American Birds, thusly: it "opens with clear flutelike note, followed by ethereal, bell-like tones, ascending and descending in no fixed order, rising until reaches dizzying vocal heights and notes fade away in silvery tinkle."

The Hermit Thrush is often heard but seldom seen, unless one visits the shadowy forest. Even then one must be quiet and patient and wait until it makes the first move. It often stays close to the forest floor, although it will sing its territorial songs from the very tip of the highest conifer. It may suddenly appear at the base of a tall conifer or among its heavy branches and sit like a sentinel until it is convinced you are not a threat. But even in the open, its colors blend in so well with the terrain that it may not be noticed at all. It has an all-brownish back, white eye rings, reddish rump and tail, and whitish underparts with black spots.

Longmire Meadow drains into the Nisqually River, which flows westward, picks up Kautz Creek from Kautz Glacier, past Sunshine Point and the park's Nisqually entrance, and eventually flows into Puget Sound. The Nisqually River is fairly typical of the numerous rivers that drain the great ice fields of Mount Rainier. Many of the park streams and rivers are milky from glacial runoff. Apparently this does not affect the aquatic wildlife, however; a number of waterbirds nest in this habitat: Harlequin Duck, Common Merganser, Spotted Sandpiper, Belted Kingfisher, and American Dipper.

The **Harlequin Duck** is unquestionably the most colorful of the five. The male is almost gaudy. It sports an overall slate blue body with chestnut stripes on its head and sides, white markings scattered here and there, including a large crescent on its face, stripes and spots on the neck, and a collar. The name "harlequin" was derived from a comical Italian character of the theater who wore a mask and multicolored tights. Females are nondescript brownish birds with white spots on their cheeks and heads. See the chapter on Olympic for additional information about this bird's fascinating life history.

This lovely creature is only an occasional summer resident at Rainier today; its numbers have drastically declined during the last half century. Mount Rainier naturalist E. A. Kitchin, in a series of 1939 nature notes, claimed that "almost any mountain stream may have a pair or family during the summer months. The faster the water the better they like it." I found the bird only once during my June visit, two females perched on a river rock at Sunshine Point. They flew upstream when I got too close.

Few birds are as well loved as the **American Dipper** or "water ouzel." While Rainier's Harlequin Duck populations have dramatically declined, probably due to impacts on the species' wintering grounds, the full-time resident dipper has apparently held its own. It is common today on all of the park's rivers and streams. I found it at Sunshine Point, Longmire, at Carter Falls, on Edith Creek at Paradise, and below Silver Falls near Ohanapecosh. On June 16 at Carter Falls, I discovered a pair of birds carrying food to an apparent nest among huge boulders just above the falls. One bird was carrying a great billful of insects, so many in fact that I wondered how it could possibly see where it was going.

Most dipper sightings are of birds standing on rocks in midstream, then flying or walking into the rapid torrent to totally disappear for several seconds, only to reappear with food. The bird will then dip up and down on bent knees, and, if close enough, one can also see it blink several times, the white eyelids closing and opening like shutters.

Kitchin provides us with a fascinating description of the dipper: "It looks and in some of his habits he might be considered a big water wren, and that is what he really is. In early spring, while the snow is still deep on both sides of the stream, we hear and are thrilled by his song, reminding us of the Winter Wren's notes, only louder. He has the same excited, jerky motions of the Winter Wren as he flits from stone to stone, and both build a round nest of green moss with the entrance on the side."

The Rampart Ridge Trail runs from Longmire Meadow for 1.8 miles to Viewpoint, then loops back to Longmire (a total of 4.7 mi.) via a portion of the Wonderland Trail. This loop route provides the hiker with an excellent perspective of the closed coniferous forest bird life. Most numerous are the Pacific-slope Flycatcher, Chestnut-backed Chickadee, Red-breasted Nuthatch, Brown Creeper, Winter Wren, Golden-crowned Kinglet, Hermit and Varied Thrushes, Dark-eyed Junco, and Pine Siskin. Fewer numbers of the Red-tailed Hawk, Blue Grouse, Vaux's and Black Swifts, Hairy Woodpecker, Northern Flicker, Hammond's Flycatcher, Steller's and Gray Jays, American Robin, and Yellow-rumped, Townsend's, and Hermit Warblers can be expected.

Townsend's and Hermit Warblers provide a real challenge for the observer; most good views during the nesting season are a matter of luck and patience. Both are lovely black-and-yellow birds that sing high overhead. To find them, you had best memorize their songs. Roger Tory Peterson includes the best descriptions in *A Field Guide to Western Birds*: Townsend's voice is "'dzeer dzeer dzeer tseetsee' or 'weazy, weazy, seesee.' The first 3 or 4 notes similar in pitch, with a wheezy, buzzy quality, followed by 2 or more high-pitched sibilant notes." The hermit warbler sings "3 high lisping notes followed by 2 abrupt lower ones: 'sweety, sweety, sweety, chup' chup" or 'seedle, seedle, seedle, chup' chup'.' Abrupt end notes distinctive."

Habitat preferences are similar, although the Townsend's Warbler prefers more open areas, whereas the Hermit's apparent preference is for dense conifer stands. Male Townsend's Warblers are most attractive, sporting coal black crowns, cheeks, throats, upper breasts, and streaked flanks, and bright yellow eyebrows, collars, and underparts; females are duller versions. Male Hermit Warblers possess all-yellow faces and foreheads, black throats, and white underparts with gray streaks; females are duller versions.

As I climbed the ridge, at least three **Blue Grouse** hooted above me. From a distance, it was difficult to determine the exact direction because of their ventriloquistic characteristic. But at the top of the ridge, I was able to walk to within 20 feet of a tree on which a grouse was calling. Even then I had to walk back and forth on the trail, triangulating the call, before I was sure. And then, once I pinpointed the location, it took me another ten minutes of searching the moss-draped fir before I located the source. It continued its hooting, transcribed by Hoffmann as "'broo, broop, broop, burroo broo broo,' increasing at first in volume, then diminishing."

The grouse's general appearance was dark brown, barred with blackish. Its breast feathers were spread to display its bulging, bare, orange-yellow skin, like a bull's-eye with white feathers tipped with black; and its orange combs were swollen and extended. The inflated neck sacs amplify the hooting. Such sounds, which may be heard for as much as 500 feet, apparently attract females; nests are built in depressions on the ground at the base of a tree, fallen log, or rock.

Viewpoint provided a marvelous view of Mount Rainier. Although clouds covered a portion of the lower left side, the entire crest was clear, with a vivid blue sky background. A pika chipped at me from the talus slope below. Several Vaux's Swifts and a pair of Black Swifts careened across the sky. Pine Siskins called "ze-e-e-e-e" from the treetops. Then, almost at my feet, a recently fledged Dark-eyed Junco was begging a concerned parent for a handout. The adult shoved a billful of food into the open mouth, but the begging continued; the youngster followed the adult when it went off to search for more insects.

Dark-eyed Juncos are one of Rainier's most abundant songbirds, present in every habitat below the tree line. Their small size, all-black hoods, pink bills and buff sides, and dark wings and tails, with white outer tail feathers, make identification easy. It seems that this little bird, earlier known as "Oregon Junco," is usually the first bird to respond to any spishing sound. It also is one of the park's most vocal songsters, often singing its musical trill from a high perch. Most sightings, however, are of birds on the ground. Hoffmann points out that "they hop actively over open ground, accompanying each movement with a partial opening of the tail which shows the white outer feathers. They do not scratch for their food like the larger sparrows but pick up the seeds from the bare surface."

Mount Rainier's closed conifer forest is at its best along the Grove of the Patriarchs Nature Trail near the Stevens Canyon entrance. One morning in June I walked the trail, watching for whatever wildlife might appear. Chickarees, chipmunks, and a fawn-colored snowshoe hare, with snow-white feet, were most obvious. Only a few birds were evident: Violet-green Swallow, Swainson's Thrush, Warbling Vireo, and Orange-crowned Warbler along the river, and Pacific-slope Flycatcher, Golden-crowned Kinglet, Brown Creeper, and Winter Wren in the forest.

Just beyond the footbridge, I encountered a family of **Winter Wrens**, tiny reddish brown birds with stubby tails and amazing energy. They never were still. The parents were actively searching for insects, which they immediately brought to their begging youngsters. I was less than 5 feet away from one baby that was shaking all over with anticipation. I watched an adult shove a long-legged fly down its throat; then, for a very brief moment, apparently still with the urge to sing, it sang a two-second ditty so low that it was barely audible. Truly a "whisper" song. Then it was again off on its search for food.

Winter Wrens normally sing a loud series of tinkling notes that go on and on. Arthur Cleveland Bent describes their song as "a rising and falling series of high-pitched notes, a fine silver thread of music lasting about seven seconds and containing 108 to 113 separate notes." The songs emanate most often from the moist, moss-draped undergrowth, but males will also sing from some high snag or branch beneath the forest canopy. Nests are hidden in cavities in rotted stumps and logs or in roots and debris. Males also build one to four "dummy" nests to fool predators.

Many of the park's lower slopes and old burns are covered with rather dense alder thickets. These habitats are the domain of the Wilson's Warbler and Fox Sparrow, although other birds that frequent snags and forest patches also are present: Rufous Hummingbird; Olive-sided Flycatcher; Varied Thrush; American Robin; Yellow-rumped, Townsend's, and Yellow Warblers; and Dark-eyed Junco. The Wilson's Warbler is a little, all-yellow songster, except for the male's solid black cap, with a rather distinct song: "chit chit chit chit CHIT CHIT CHIT." To my ear, its song seems to end just when it gets fully started.

Fox Sparrows were numerous among the alders at the old burn near Bench Lake, in upper Stevens Canyon. As many as five were singing at one time. I had several good looks from the Bench Lake Trail, including one individual which, judging from the huge billful of materials it was carrying, was apparently nest-building. Rainier's Fox Sparrow possesses a dark gray hood and back, large yellowish bill, and dark brown, arrow-shaped breast spots that form a heavy stickpin. It often sings from a high post, such as a snag or higher shrub, and its loud, ringing, and melodic song is one of the mountain's finest. Hoffmann states that its song "varies greatly, even in the same individual, but once heard is unmistakable; it includes as one of the opening phrases a pair of loud sweet notes, 'swee chew' or 'wee chee' followed by trills and runs."

AT AND ABOVE THE TREE LINE

Paradise and Sunrise visitor centers were built at the upper edge of the subalpine forest, just below the tree line and the marvelous meadows and snowfields of the alpine tundra. It is impossible to obtain a true perspective of Mount Rainier National Park without visiting one of these areas. Four Corvids (crow family) are sure to be part of that experience: Steller's and Gray Jays, Clark's Nutcracker, and the Common Raven. These gregarious species are especially interested in any handouts that might be available, despite the fact that such snacks are forbidden because they are not as healthy as their natural foods. If a handout is not readily available, they may then inspect the front of your vehicle for any smashed insects that you may have inadvertently brought along.

Steller's Jays are easily identified by their all-blue plumage and tall blackish crests; the noncrested Gray Jays possess soft gray plumage with black foreheads; and the larger Clark's Nutcrackers possess all-gray bodies with black wings with white patches, and black feet, eyes, and bills. In flight, the white trailing edges of their wings and broad, white outer tail feathers are conspicuous. Their almost continuous call, a guttural and drawn-out "kra-a-a," is often heard even before the bird appears.

Clark's Nutcrackers are, without a doubt, one of the bird world's most fascinating members. They not only thrill park visitors with their aerial acrobatics, boldness, and apparent curiosity, but they also possess an exceptionally keen memory. Nutcrackers are able to remember the location of about one thousand seed caches from one season to another. See the chapter on Crater Lake for details on how they are able to find their stashes.

Gray Jays also cache food for the winter. But rather than hiding food underground like their larger cousins, they hide their food on conifers. The partially digested materials, coated with sticky fluids from their mouths, are literally glued in place. This system of storage permits them to remain in their territories throughout the winter months when little fresh food remains. It also allows them to nest in late winter, when the snow is still deep.

Most often, Gray Jays seem to appear out of nowhere, sailing through the conifers like aerial ghosts. They seldom flap their wings, except to get off the ground. Mostly gray with white foreheads and blackish napes, they can easily remind one of soft, stuffed toys. Their calls are usually little more than soft "whee-oo" notes, but they can be loud and boisterous like their blue-crested cousins.

Other birds to be expected in the subalpine zone include the Violet-green Swallow, Barn Swallow (in the vicinity of the buildings), Mountain Chickadee, Ruby-crowned Kinglet, Hermit and Varied Thrushes, Yellow-rumped Warbler, Dark-eyed Junco, and Pine Siskin.

Above the tree line, as soon as the ground becomes snow free, wildflowers swiftly dominate the scene. The Paradise and Sunrise trails provide ready access to the high alpine meadows. The Skyline Trail, a 5-mile loop from Paradise, is one of the best. Four alpine birds are possible along this route: the White-tailed Ptarmigan, Horned Lark, American Pipit, and Gray-crowned Rosy-Finch.

The **White-tailed Ptarmigan** is not an easy bird to find; it blends in so well with the flowering landscape that locating one is often serendipitous. Ptarmigans are a true alpine species, changing their plumage with the seasons. In summer, they possess a mottled plumage of blacks, browns, golds, and whites, along with a scarlet eye comb above each eye and a snow-white belly and wings. In winter they are all white, except for their scarlet combs and black bills.

Ptarmigans, sometimes called "snow quail" or "snow partridge," live on the snowfields year-round, except during the harshest winters when they may move into protected valleys. Besides their winter plum-

Fig. 9. White-tailed Ptarmigan

age change, they also grow "snowshoes," dense mats of stiffened feathers on their toes that help in walking over fresh snow. Traveling ptarmigans have been seen descending rocky slopes by sliding with legs forward and tail spread behind and used as a rudder. In addition, their nostrils are closed by dense feathers that keep out winter snow. And Terres reports that ptarmigans "fly directly into soft snowbanks to sleep; dozens may roost close together, but none walks to the roosting places because their tracks could be followed by weasels, foxes, lynxes, or other predators which might catch ptarmigans in their sleep." They feed on twig tips and buds that protrude above the snow in winter, and add flowers, leaves, and insects to their summertime diet.

Watch for **Gray-crowned Rosy-Finches** at rocky places along the edges of snowfields. While ptarmigans are full-time park residents, rosy-finches spend only the summer months in the park, migrating east into the warmer valleys for the winter. Kitchin reports that rosy-finches arrive early and follow the receding snow up the mountain-side: "By the time patches of bare earth appear at Paradise Valley these birds are busy feeding young. A pair may be seen on one of these bare spots picking up seeds and buds of the heather . . . then off they go, over a precipice and across a glacier bed to some niche on the perpen-dicular face of a rock to feed their brood."

"In the fall," Kitchin continues, "these birds gather in large flocks and fly about, over the bare ground and talus slopes, in action much like a flock of blackbirds over a stubble field. They commence their slow migration down the eastern side of the Cascades, stopping at times for several days in one locality." They will often roost together in tight-knit groups, and winter roosts of up to one thousand individuals have been reported in abandoned cliff swallow nests, cave entrances, mine shafts, and other artificial structures.

Winter in the high country is very different; birds are few and far be-tween. Only a few of the hardiest species remain: Blue Grouse, White-tailed Ptarmigan, Northern Pygmy-Owl, Northern Flicker, Hairy Woodpecker, Steller's and Gray Jays, Clark's Nutcracker, Common Raven, Mountain Chickadee, Red-breasted Nuthatch, Winter Wren, Dark-eyed Junco, and Pine Siskin.

In summary, Mount Rainier's bird checklist includes 166 species, of which 109 are listed for summer, and assumedly nest. Of those 109 species, 7 are waterbirds, 15 are hawks and owls, and 9 are warblers. Only one bird, the Snow Bunting, is listed (rare) only in winter.

BIRDS OF SPECIAL INTEREST

Harlequin Duck. Watch for this multicolored duck at swift streams in the highlands during spring; duller females and juveniles remain throughout the summer.

Blue Grouse. Courting males give hooting calls in spring, often from concealed perches on tall conifers.

White-tailed Ptarmigan. Watch for this brown, black, and gold bird, with white wings and tail, on the alpine tundra.

Vaux's Swift. This fast-flying, grayish bird looks like a flying cigar with long, narrow, and stiff wings.

Steller's Jay. One of Rainier's most familiar birds, this blue bird possesses a tall, blackish crest and an obnoxious personality.

Gray Jay. This is the noncrested, gray-to-whitish jay of the forest; it is best known as "camp robber" or "whiskey Jack."

Clark's Nutcracker. Most common at Paradise and Sunrise, it sports an all-gray body with black-and-white wings, with white trailing edges.

Chestnut-backed Chickadee. This common chickadee has a chestnut back and sides and sings a drawling "tsic tsic tsic tyee" song.

American Dipper. Watch for this plump, all-gray-brown bird at swift streams; it actually feeds on aquatic insects obtained underwater.

Winter Wren. A tiny, reddish brown bird of the dense forest, which is heard more often than seen; it sings an extensive tinkling song.

Varied Thrush. This robin-sized bird, with orange and brown plumage, sings an eerie, bell-like whistle that is soon repeated on a higher or lower pitch.

Hermit Thrush. A bird of the shadowy forest, it sings wonderful flute-like songs from the high treetops in spring.

Gray-crowned Rosy-Finch. Watch for this rosy and gray finch at the edges of snowfields during the summer months.

Dark-eyed Junco. This little "snowbird" has an all-black hood, buff sides, and black tail with white outer tail feathers.

Fox Sparrow. Common in summer on alder-covered slopes, it is a large dark gray sparrow that sings a loud, ringing and melodic song.

Crater Lake National Park, Oregon

Clark's Nutcrackers are unquestionably the park's most obvious wildlife. These medium-sized, gray, black, and white birds, with conspicuous white trailing edges to their black wings, are widespread along the Rim, perched on various snags or on the low stone walls along the parking areas and sidewalks, or flying by. If they are not immediately apparent, their guttural and drawn-out "kr-a-a" or "chaar" calls are commonplace. It is doubtful that any visitor to this marvelous park of the extraordinary deep blue lake will miss Clark's Nutcrackers.

A member of the crow family (Corvidae), the nutcracker is one of the bird world's most fascinating members. It is an amiable and personable bird that seems almost to seek out human companionship. It also is among our smartest birds, able to remember the location of about one thousand seed caches from one season to another.

In the fall, when conifers ripen, nutcrackers pry seeds from the cones in crowbar fashion with their sharp, heavy bills, then hide the seeds on south-facing slopes for winter use. They possess a special sublingual pouch under their tongues in which they are able to carry up to ninety-five seeds per trip. A study by ornithologist Stephen Vander Well, of Utah State University, proved that nutcrackers were able to recall where they had cached their seeds in relation to key landmarks, such as rocks. When those key landmarks were moved, the areas the birds searched were displaced to an equivalent degree.

Fig. 10. Clark's Nutcracker

This food-caching behavior is not unique to Clark's Nutcrackers. Several other birds also store food, including woodpeckers, jays, chickadees, nuthatches, titmice, shrikes, and a few hawks and owls. But only nutcrackers are able to cache food in such large quantities. Steller's and Gray Jays store lesser amounts of seeds in the ground, carrying the seeds in an expandable esophagus. Nutcrackers bury their caches at a perfect depth for germination; thus seeds that go undiscovered (approximately 30 percent) can produce more trees. Paul Ehrlich and colleagues, in *The Birder's Handbook*, write that "pines appear to have evolved cone and seed structures and fruiting times that increase the chance for their seeds to be buried (planted) by the birds."

The scientific name for Clark's Nutcracker is *Nucifraga columbiana:* the genus name is Latin for "nut-breaker," and the species name was derived from the Columbia River, the place where William Clark, of the Lewis and Clark Expedition, was the first to collect one in about 1804. Its common name comes from its discoverer.

THE PARK ENVIRONMENT

Crater Lake has no equal! No other lake possesses such deep blue color and majesty. No other lake combines such incredible depths (the world's seventh deepest at 1,932 ft.) with such striking cliffs and the contrasting greenery of the surrounding landscape. No other lake seems to so perfectly illustrate Henry David Thoreau's description: "A lake is the landscape's most beautiful and expressive feature. It is earth's eye, looking into which the beholder measures the depth of his own nature."

By national park standards Crater Lake National Park is not large, at 183,224 acres, but it is one of North America's most exquisite jewels. Primarily a summertime park, the Rim, with an average elevation of about 7,000 feet above sea level, normally lies beneath 15 feet of snow all winter. By mid-July, however, displays of wildflowers dominate the open meadows and roadsides. The 33-mile Rim Drive (usually open by July 1), which completely circles the lake, provides easy access to these high-country meadows, numerous viewpoints, and a 1-mile trail to the Cleetwood Cove boat dock.

Crater Lake itself is the product of a violent volcanic eruption about 7,700 years ago, when 12,000-foot-high Mount Mazama expelled great amounts of pumice and ash around the mountain and across vast stretches of North America. There is no evidence that the mountain exploded outward, but with so much of its innards gone, the mountain collapsed downward (or inward upon itself), leaving a great caldera or basin, with a maximum width of 6 miles, that eventually was filled with water from rain and snow to form Crater Lake.

Today the park's vegetation can be divided into six rather distinct communities. A subalpine environment, dominated by wind-blown whitebark pine, mountain hemlock, Shasta red fir, and open pumice fields, generally lies above 7,000 feet. Heavy stands of mountain hemlock, intermixed with Shasta red and noble firs and scattered western white and lodgepole pines, occur in protected areas along the Rim and down to about 5,500 feet. Below these communities, usually in dry, well-drained areas, is a lodgepole pine forest. The surrounding flatlands, especially to the south, are often dominated by open ponderosa pine communities, with scattered Douglas-fir and white fir. Chaparral, consisting primarily of greenleaf manzanita, bittercherry, and snowbush ceanothus, covers some of the drier, southern slopes. Finally, riparian vegetation, composed primarily of willows, alders, and black cottonwood, hugs the streamsides.

The National Park Service operates visitor centers at park headquarters and Rim Village. Each offers information services, exhibits, and a sales outlet; bird field guides, a checklist, and a useful booklet by Dick Follett—*Birds of Crater Lake National Park*—are available.

The visitor centers, along with the Sinnott Memorial Overlook, serve as hubs for the park's varied interpretive activities in summer. These range from various talks and walks to evening programs at Mazama Campground. Schedules are posted on bulletin boards and are available at the entrance stations and visitor centers.

Additional information can be obtained from the superintendent, Crater Lake National Park, P.O. Box 7, Crater Lake, OR 97604-0007; (541) 594-2211.

BIRD LIFE

Clark's Nutcracker is but one of six Corvids that have been found within the park. The largest of these is the all-black Common Raven, easily identified by its size, huge bill, and wedge-shaped tail. The similar but smaller American Crow, as well as the long-tailed Black-billed Magpie, are rare park visitors only. The other common Corvids include Gray and Steller's Jays.

Ravens commonly soar over the Rim, using thermals that rise along the slopes just like Turkey Vultures and hawks. At times, especially in very early spring when courting, they put on a wonderful show of aerial adroitness. One day in June I watched a pair of ravens in an obvious courtship flight high over Rim Village. Their chase included some amazing twists and turns, tumbling, spirals, rolls, and dives, and then sharp ascents, sometimes while flying parallel to one another, and all the while they called hoarse "caaw" notes, along with a variety of gurgling, croaks, and clucking sounds. Ravens actually nest on the high cliffs surrounding the lake, and they often can be found patrolling the park roads during the early morning hours, looking for roadkill from the previous night.

Gray Jays are also common along the Rim but are more subtle than their large, black cousins. These gray-and-white birds usually appear like soft, fluffy ghosts, sailing out of the forest or from tree to tree in search of a handout. They seem to be entirely without fear and filled with curiosity, landing on a picnic table or investigating one's presence wherever one might go. They seem to follow hikers, chattering in low tones all the while.

Widely known as "camp robbers," Gray Jays also store food for the winter. But rather than hiding it underground like the longer-billed nutcrackers, they store their booty on conifers. The partially digested food is coated with sticky fluids from their mouth glands that help it to stick in place. This storage system permits them to remain on their territories throughout the winter months when little fresh food remains. It also allows Gray Jays to nest in late winter, when snow is still deep, before the arrival and/or growth of a new food supply.

Steller's Jays are also present in the park in summer, especially in areas of human use. But most of these birds move out of the park into lower, warmer valleys for the winter months. This is the royal blue jay with a high, blackish crest and a wide variety of vocalizations. Arthur Cleveland Bent, in his Life History series, reports that Steller's Jays possess "low-pitched raucous squawks, different from other kinds of jays; calls harsh 'waah, waah, shaack, schaak, schaak,' and mellow 'klook, klook, klook,' and shrill hawklike cries: 'kweesch, kweesch, kweesch'; has sweet soft song and somewhat like 'whisper song' of robin; female has a rolling click call; is superb at imitating scream of red-tailed hawk."

The entire Rim Drive has numerous viewpoints that are also excellent places to observe soaring birds. Most common are the Turkey Vulture, Red-tailed Hawk, and, in late summer, American Kestrel. The Osprey; Bald and Golden Eagles; Northern Harrier; Sharp-shinned, Cooper's, Swainson's, Ferruginous, and Rough-legged Hawks; Merlin; and Prairie and Peregrine Falcons are also possible.

The **American Kestrel** is the smallest of the falcons and also one of the most colorful. Males sport a rust-colored back and tail with contrasting bluish wings and white cheeks with double black moustache stripes. They also possess a distinct call: a loud and shrill "killy killy killy." Their flight is normally rapid and direct. William Clark and Brian Wheeler, in A Field Guide to Hawks of North America, point out that it "soars on flat wings, often with the tail fanned. . . . The American Kestrel is the only N. American falcon to hunt regularly by hovering (wings flapping) or, in strong wind, by kiting (wings held steady)."

Kestrels, earlier known as "Sparrow Hawks," "breed in small numbers at lower elevations, particularly in the southern portion of the park, and is a common late summer visitor at higher elevations," according to Follett. Grasshoppers are an important food item at that time of year, although kestrels are known to feed on a wide variety of prey, including numerous insects, amphibians, reptiles, and small birds and mammals.

Peregrine Falcons are also present in summer, but seeing one of these marvelous creatures is largely a matter of luck; they can occur anywhere along the Rim. The park's resource management specialist, Mac Brock, told me that only one pair is known to nest in the park. He also said that surveyors found a total of eleven pairs of Spotted Owls. Both of these raptors are listed as "endangered" by the U.S. Fish and Wildlife Service.

The Rim Drive is a must for anyone visiting Crater Lake; lake views are spectacular and the drive provides a smorgasbord of high-country habitats. One morning in early July, I recorded more than two dozen bird species along this route. Most frequent, after the Gray Jay, Common Raven, and Clark's Nutcracker, were the brightly colored Rufous Hummingbirds; Northern (Red-shafted) Flickers; little black-and-white Mountain Chickadees; Red-breasted Nuthatches, calling nasal "nyak nyak nyak" notes from the forest; red-breasted American Robins; Hermit Thrushes, identified by their lovely, flutelike songs; Yellow-rumped Warblers, flycatching from various snags; Western Tanagers, with the male's black, yellow, and red plumage; Dark-eyed Juncos, calling high-pitched trills from the treetops; Cassin's Finches, sitting on the tops of various trees or flying by; and dozens of tiny Pine Siskins, usually detected first by their rising "schreeee" notes.

Fewer numbers of the Turkey Vulture, Red-tailed Hawk, Black-backed Woodpecker, Olive-sided Flycatcher, Violet-green Swallow, Steller's Jay, Brown Creeper, Rock Wren, Golden-crowned and Ruby-crowned Kinglets, Mountain Bluebird, Townsend's Solitaire, American Pipit, Chipping Sparrow, Red Crossbill, and Evening Grosbeak were detected.

Yellow-rumped Warblers were especially common that day; on several occasions I was able to observe them at eye level from the elevated roadway. Gorgeous little birds, Yellow-rumps sport a contrasting black, yellow, and white plumage with five yellow spots: on their caps, throats, sides, and rumps. Follett regards this bird as "Crater Lake's most abundant breeding bird," in part due to its feeding diversity:

> An energetic, active species, the Yellow-rumped Warbler feeds in a variety of niches: by moving along a limb or branch, fluttering at the end of a branch, hanging upside-down on the underside of a branch, flycatching, and feeding on the ground. This versatile warbler is also a very hardy bird which is able to tolerate greater temperature extremes than most other warblers. This species arrives in Crater Lake in April, and many do not depart until November.

Dark-eyed Juncos were equally abundant, and also evident at all heights in the forest. However, these little sparrows spend considerable time on the ground and often can be found hopping along the edges of roads or parking areas, searching for seeds or tiny insects. Juncos are easily identified by their all-black hoods, buff backs, whitish underparts, and dark tails with broad white edges. Nesting juncos prefer meadow edges and open forest, constructing their nests on the ground, usually under a patch of grass or rock. Donald Farner, in an out-of-print book, *The Birds of Crater Lake National Park*, claims that nesting "begins commonly in June, although the appearance of flocks of juveniles during the first week in July indicates that a considerable amount of nesting activity begins in May. Doubtless two broods are the rule."

One of the most appealing of all the forest birds is the personable little **Mountain Chickadee**. A highly vocal and gregarious bird, it seems always to be present, either in pairs, family groups, or, by late summer, as the nucleus of flocking forest birds. Usually detected first by its rather distinct, hoarse "chick-a-dee" or "dee-dee-dee" songs, it sports white cheeks and a black bib and cap that is divided by white eyebrows. Follett describes its feeding behavior: "Chickadees survey the countless cracks and crevices of tree trunks and branches in search of spiders and spider eggs, plant lice, scale insects, butterfly and moth eggs, saw flies, leaf hoppers, insect borers and ants. In their incessant search for insect prey they often hang upside down, sometimes from the very tip of conifer branches."

The park's subalpine communities are scattered along the Rim proper as well as on the adjacent higher points, such as Garfield and Hillman Peaks, Llao Rock, and Mount Scott. The extensive open slopes are good places to find the Prairie Falcon, Horned Lark, Rock Wren, and Gray-crowned Rosy-Finch in summer. Prairie Falcons, pale brown to gray birds with pointed wings and dark auxiliaries (wing pits), nest on the caldera walls and hunt this open terrain. Horned Larks, sparrow-sized birds with a black-and-yellow head pattern, with black hornlike protrusions, nest in these areas; they perform wonderful courtship flights. Rock Wrens, grayish birds with a habit of bobbing up and down, frequent the rocky slopes; they are readily located by their ringing calls and songs: sharp "tick-ear" calls and songs that possess all or part of a full song that John Terres, in the *Audubon Society Encyclopedia of North American Birds*, describes as "keree keree keree, chair chair chair, deedle deedle deedle, tur tur tur, keree keree trrrrr." American Pipits are faintly streaked birds with dark brown backs that walk about the grassy terrain with considerable teetering motions and have a "pit" or "chip-it" call.

Gray-crowned Rosy-Finches, one of the park's most fascinating alpine finches, possess a rather unique plumage: black forehead, gray nape and cheeks, blackish back and chest, and pink underparts, rump, and wings. Follett points out that they possess "special adaptations which allow them to live in the harshest of land environments. Their heart and respiratory systems are more highly developed than those of species which occur at lower elevations." Watch for this lovely creature at the edges of snowbanks; it feeds on seeds gleaned from plants emerging from the snow or insects blown onto the ice. The Garfield Peak Trail, behind Crater Lake Lodge, is a good place to find this bird in May to early July. Afterward, rosy-finch flocks, which fly as a unit like blackbirds wheeling over a cornfield, seek out sheltered areas along the Rim or on the north side of Mount Scott.

LAKE BIRDS

The best way to find lake birds is via a boat tour from Cleetwood Cove. However, the **Spotted Sandpiper** is the only waterbird that consistently nests there. It frequents the grassy shoreline and can often be seen flying close to the water with rapid, stiff wingbeats and distinct "weep" calls. Summer birds possess heavily spotted breasts and yellowish bills and legs, and have a habit of teetering. The Spotted Sandpiper's presence inside the Crater Lake caldera supports the commonly held belief that this species may be North America's most widespread shorebird; its unusual breeding strategy may be responsible. The species practices polyandry. The females arrive on their nesting grounds first and establish territories before the arrival of the males. Once the males arrive, the polyandrous females will actually form their own harem, occasionally competing with other females in fierce combat.

Multiple males permit the combative female to lay up to five clutches of four eggs, which are then cared for by her male consorts.

Crater Lake's other common waterbird is the California Gull, a summer visitor only, probably from breeding colonies at Upper Klamath Lake. Adult gulls sport gray mantles with black-and-white wing tips, white heads and tails, yellow-green feet, and yellow bills with black and red spots. Immatures' upperparts vary from all mottled to dark; they have whitish underparts, with black to two-tone bills.

During four visits to the lake in one summer, I also recorded the American White Pelican, Double-crested Cormorant, Canvasback, Common Merganser, Killdeer, Ring-billed Gull, and American Coot. Other possible summer-visitor waterbirds include the Common Loon, Eared Grebe, Great Blue Heron, Mallard, Northern Pintail, Wood Duck, Lesser Scaup, Barrow's Goldeneye, Ruddy Duck, and Hooded Merganser. Also watch for Violet-green, Northern Rough-winged, and Barn Swallows fly-catching over the lake.

LOWLAND FORESTS AND STREAMS

The extensive forest communities below the Rim support a more extensive bird life in summer, and Mazama Campground and the adjacent 1.7-mile Annie Creek Canyon Trail provide easy access. Common campground birds include the Northern Flicker, Western Wood-Pewee, Gray and Steller's Jays, Mountain Chickadee, Red-breasted Nuthatch, American Robin, Yellow-rumped Warbler, Western Tanager, Dark-eyed Junco, Chipping Sparrow, Cassin's Finch, and Pine Siskin. And another two dozen or more species are possible along the nature trail proper.

Annie Creek and streamsides also support a number of interesting birds, none more exciting than the **American Dipper.** This plump, slate gray bird with a stubby tail actually forages underwater for water insects and larvae. Our only truly aquatic songbirds, dippers (sometimes called "water ouzels") swim through the swift currents and walk on the stream bottoms. The bird's name comes from its habit of dipping up and down with its whole body. Dipper nests are bulky, oven-shaped structures, usually a foot in diameter and constructed under

Fig. 11. American Dipper

waterfalls, on ledges, or among exposed tree roots. Females, like Spotted Sandpipers, are polygamous, mating with several males and defending their mates against other females. Young of the year are very precocious and can climb, dive, and swim on departing the nest. See the chapter on Mount Rainier for details on the special adaptations that support their aquatic behavior.

Farner includes a fascinating account of dippers feeding along Munson Creek in winter where the "snow was about eight feet deep and approximately two-thirds of the stream was covered with snow." He continues:

> At irregular intervals, there were open "wells" down to the surface of the stream which had no ice on it. These wells were three to six feet in width, about eight feet deep, and from three to 30 feet in length. The distance between the bottom of the snow between the "wells" and the surface of the water appeared to be six inches to two feet. For nearly two hours, I observed the habits of a Dipper in this situation. I discovered that it was unnecessary for it to fly from "well" to "well" but rather that it frequently traveled from one to another beneath the snow. From the fact that I ob-

served it taking food as it left the "well" to move along the stream beneath the snow, and from the time required to pass from "well" to "well," I became quite convinced that it fed along the sections of the stream beneath the snow as well as along the parts exposed in the bottom of the "well." The maximum distance traveled by the Dipper under the snow was about 30 feet.

Other streamside birds in summer include the tiny Calliope Hummingbird; Red-breasted Sapsucker, with the male's all-red breast and head; Black-capped Chickadee, like the Mountain Chickadee but without the white eyelines, and with a slower "chick-a-dee-dee-dee" song; Swainson's Thrush, which sings an upward-rolling "wip-poor-wil-wil-eez-zee-zee" song; Warbling Vireo, which sings a husky, rambling warble all summer long; Orange-crowned Warbler, a nondescript, yellowish bird; MacGillivray's Warbler, distinguished by its all-gray hood with broken, white eye rings; Wilson's Warbler, with the male's black cap and yellow body; and Lincoln's Sparrow, a perky little sparrow with a buff chest.

Several additional forest and forest-edge habitat birds can be found along the Annie Creek Canyon Trail: Blue Grouse, Northern Flicker, Williamson's Sapsucker, Pacific-slope and Olive-sided Flycatchers, Violet-green Swallow, Steller's Jay, Red-breasted Nuthatch, Winter Wren, American Robin, Varied and Hermit Thrushes, Western Bluebird, Townsend's Solitaire, Golden-crowned and Ruby-crowned Kinglets, Cassin's Vireo, Nashville and Yellow-rumped Warblers, Western Tanager, Chipping Sparrow, and Pine Siskin.

During early summer mornings, when the birds are most active, their varied songs are all prevailing. Thrush songs can dominate the environment, with the **Hermit Thrush** leading the choir. This lovely forest thrush is more often heard than seen, except when it is singing from the topmost branches of a high conifer. Then its spotted underparts and brownish back, with a reddish rump and tail, can be studied for a considerable time. In the undergrowth, its colors blend in so perfectly with the shadows that it is next to impossible to find.

The uncommon Varied Thrush sings a very different song, like long, quavering whistles, followed by a pause, then another quavering whistle at a higher or lower range. The Townsend's Solitaire sings a song with a prolonged, melodious series of rapid warbling notes that are sweet, clear, and rich in their delivery. But the most continuous songs, especially during the early morning and evening hours, are those of the American Robin. This bird's sweet caroling has often been described as "cheerily-cheery-cheerily-cheery."

Below the Annie Springs area to the south are scattered stands of ponderosa pine and hillsides dominated by chaparral vegetation. These two habitats support a few additional birds, normally not found elsewhere. The ponderosa pine forest species include the Ruffed Grouse; Common Nighthawk; Pileated, Hairy, and White-headed Woodpeckers; Hammond's Flycatcher; White-breasted and Pygmy Nuthatches; and Evening Grosbeak. Chaparral birds include the Mountain Quail (rare), Dusky Flycatcher, Green-tailed Towhee, and Fox Sparrow.

Winter comes early to the Crater Lake high country, forcing many of the birds to lower, warmer elevations. Most of the Neotropical species leave their breeding grounds as early as late July or August. Others linger behind so long as they can find adequate food. But by late October, only a few of the hardiest species remain: Prairie Falcon, Blue and Ruffed Grouse, owls, woodpeckers, Corvids, chickadees, nuthatches, Brown Creeper, dipper, Golden-crowned Kinglet, juncos, and a few finches.

In summary, the Crater Lake checklist of birds includes 192 species. Only 30 of the 192 species are shown to nest (undoubtedly incorrect): none are waterbirds, four are hawks and owls, and none are warblers.

BIRDS OF SPECIAL INTEREST

American Kestrel. Watch for this little falcon hunting insects in open areas along the Rim Drive during late summer.

Spotted Sandpiper. This is the spot-breasted shorebird that flies with shallow, stiff wingbeats low over the lake; it usually gives loud "weep" call-notes.

Gray Jay. Common along the Rim Drive, it sports soft gray to white plumage and has a friendly manner.

Steller's Jay. This is the blue jay with a high, blackish crest; it is most common in forest areas below the Rim.

Clark's Nutcracker. One of the park's most obvious birds, it has a gray head, black-and-white wings and tail, and a loud guttural call: "kr-a-a" or "chaar."

Common Raven. Often seen flying over the Rim, it is easily identified by its all-black plumage, large size, and wedge-shaped tail.

Mountain Chickadee. This chickadee resides among the conifers; it has a black bib and head, with white eyelines.

American Dipper. Watch for this plump, dark gray bird along the streams; it actually dives underwater to forage for insects on the stream bottom.

Hermit Thrush. One of the park's finest singers, this brownish bird, with a reddish rump and tail and spotted chest, is a common summer resident of the forest communities.

Yellow-rumped Warbler. One of the park's most abundant forest birds, it has a contrasting black chest and five yellow spots: on its cap, throat, sides, and rump.

Dark-eyed Junco. This is the little black-hooded bird with noticeable white feathers on the edge of its dark tail; it sings a high-pitched trill in summer.

Gray-crowned Rosy-Finch. Watch for this pink, black, and gray finch at the edges of snowbanks; flocks often fly together like wheeling blackbirds.

Woodpeckers are an important part of the old-growth forest at Oregon Caves. When walking the park trails, it is virtually impossible not to see, hear, or find evidence of woodpeckers. Drumming and calling woodpeckers are a constant. And a surprising number of the park's abundant tree snags contain holes where these birds can often be found either searching for insects or excavating cavities for nesting. Although most visitors come to see the superb cave formations, those who hike the park trails obtain a first-hand lesson in forest ecology.

A total of seven woodpecker species have been found within the monument boundary: Lewis's, White-headed, Downy, Hairy, and Pileated Woodpeckers; Northern Flicker; and Red-breasted Sapsucker. Most abundant of these are the "Red-shafted" Flicker and Hairy and Pileated Woodpeckers.

The massive holes in the old tree trunks, both standing and on the ground, are largely the work of the Pileated, the largest of the woodpeckers by far. A lucky visitor may even watch this wonderful woodpecker at work; it frequents the forest near the park facilities and cave entrance. The **Pileated Woodpecker,** whose name is pronounced either "pi-le-at-ed" or "pil-le-at-ed," according to *Webster's Ninth New Collegiate Dictionary,* is easy to identify because of its large size and black back, tail, and underparts, white stripe that runs across its cheek and down its neck, and bright red crest. Males also possess a red patch be-

Fig. 12. Pileated Woodpecker

hind the bill. In flight, which is very bat-like, they show huge white patches below and smaller patches on top of the wings.

Despite its large size, this woodpecker is more often heard than seen. It has a loud, far-reaching "wuk wuk wuk" calls, which may sound like the "wick" notes of a flicker. John Terres reports that it also has "a loud 'high call' cuk, cuk, cuk, cuk, cuk, higher-pitched in female, the main breeding note that expresses dominance in its area." He also states that "both sexes drum, females less than males, usually on resonant place on dead tree or dead stub, a rolling tattoo lasting 3–5 seconds; drumming advertises territory and attracts a mate."

The Pileated Woodpecker, the model for the "Woody Woodpecker" cartoon character, requires considerable acreage of mature conifers to survive. These birds have disappeared in areas that have been heavily timbered, although there also are places where they have returned after reforestation. They normally are full-time residents and maintain mates year-round. Paul Ehrlich and colleagues, in *The Birder's Handbook,* summarize their nesting activities thusly: "Male roosts in nest cavity prior to egg laying; at other times roosts in nest cavity of previous year. Male incubates at night. Male and female brood up to 10 days." They also include a record of a female Pileated that retrieved her eggs from a fallen nest tree and carried them in her bill, one by one, to a new site. Nesting was resumed, and the nestlings eventually fledged.

THE PARK ENVIRONMENT

Oregon Caves was proclaimed a national monument in 1909 to protect the beautiful, 3-mile-long solution cave, earlier referred to by Joaquin Miller (poet of the Sierras) as the "Marble Halls of Oregon." Surrounded by Siskiyou National Forest lands, the 484-acre monument also contains a remnant of original forest. And the approximately 6 miles of trails provide easy access.

Located on the western slopes of the Siskiyou Mountains in southwestern Oregon, the monument is situated in a rocky canyon surrounded by steep forested slopes. Area vegetation, studied by James Agee and colleagues in developing a fire history for the area, includes seven community types: a white fir/herb community, dominated by white fir and incense-cedar, with a shrub layer of California hazelnut and wild rose; a mesic white fir/Douglas-fir community of white fir and Douglas-fir, and a shrub layer of currants/gooseberry and hazelnut; a dry white fir/Douglas-fir community of white fir, Douglas-fir, and bigleaf maple; a Douglas-fir/oak community, dominated by Douglas-fir, white fir, canyon live oak, tanoak, Pacific madrone, and incense-cedar; an Oregon white oak community, dominated by Oregon white oak and incense-cedar; a meadow community, dominated by herbs; and an alder community, dominated by Sitka alder.

A multiagency Illinois Valley Information Center, located on Highway 46 in Cave Junction, offers information services, an orientation video program, and a sales outlet; bird field guides are available. A bird checklist is also available at the monument ranger station at the entrance parking area. Interpretive activities include walks and evening talks during the summer months, guided cave tours, and the self-guided Cliff Nature Trail.

Additional information can be obtained from the area manager, Oregon Caves National Monument, 19000 Caves Highway, Cave Junction, OR 97523; (541) 592-2100.

BIRD LIFE

The most numerous woodpecker in the Oregon Caves forest is the **Hairy Woodpecker.** Only half as large as the Pileated, the Hairy Woodpecker is best identified by its white back and contrasting black wings and tail. It also has a black cap, white eyeline, black cheeks, and white nape, and males sport a red crown stripe. Hairy Woodpeckers are mixed-forest birds, while the smaller look-alike Downy Woodpeckers prefer broadleaf communities, such as alder thickets. Besides their overall size differences, bill sizes are an even more dependable method of identification; a Downy Woodpecker's bill is tiny by comparison.

All the resident woodpeckers construct nest holes in live or dead trees and snags almost every year, producing lots of unused nest cavities. A number of other forest birds take advantage of the vacated cavities for nests: Western Screech-Owl, Northern Pygmy-Owl, Northern Saw-whet Owl, Violet-green Swallow, Mountain and Chestnut-backed Chickadees, White-breasted and Red-breasted Nuthatches, and Winter

Wren. Each of these variable-sized birds is able to find the right-sized cavity.

Although owls are year-round residents in the forest, finding one is usually a matter of chance, even when knowing when and where to look. Locating songbirds is much easier; nesting birds readily advertise their locations during the daytime, especially during early morning, and a few sing all day.

The Cliff Nature Trail provides the visitor with a good introduction to the area's bird life. A dozen or more species can easily be found along this trail. If you have not already met the **Steller's Jay** at the monument parking area, along the entrance road, or at the cave entrance, it surely will be one of the first birds encountered on the trail. This is the large, crested blue jay with a blackish nape and crest, a loud voice, and an obnoxious personality. It apparently thinks it owns the premises, and will continuously monitor your activities, especially if it thinks you are about to have lunch or a snack. A member of the crow (Corvidae) family, it has an amazing capacity to carry away and store food in caches hidden on trees or on the ground. It may even steal food from other birds' caches.

Ralph Hoffmann, in his delightful book, *Birds of the Pacific States*, provides us with a description of the personality of the Steller's Jay:

> The Crested Jay is a noisy, inquisitive bird, hopping about a camp in search of bits of food, or screaming an alarm at our approach. It is one of the first birds to discover a hawk perched in a tree, or an owl in its hiding place, and to proclaim its discovery with angry cries. Besides the ringing 'tchek' . . . generally given in flight, the Crested Jay utters from its perch a loud 'kweesch, kweesch, kweesch.' It has besides a deeper 'chu-chu-chu' and a note resembling a squeaking wheelbarrow, 'kee-lu, kee-lu.' It also utters screams so like those of the Red-tailed and Red-bellied Hawks as to deceive the listener. Occasionally from the cover of dense foliage, it utters a formless succession of liquid, pleasing notes quite unlike its usual discordant notes, or a purring or rolling note.

Other obvious birds to be found along the Cliff Nature Trail include the Dark-eyed Junco, Pacific-slope Flycatcher, Red-breasted Nuthatch, Chestnut-backed Chickadee, and Winter Wren, more or less in that order. **Dark-eyed Juncos** are usually abundant, although they may be secretive near their nests. This is the little black-headed bird with pinkish sides and white edges on its dark tail, especially evident in flight. Once known as "Oregon Junco," it has been lumped with the Rocky Mountain and eastern forms under the generic "dark-eyed" name. Juncos also are known as "snowbirds" in the southern parts of their range where they overwinter. Oregon Cave's junco population, however, seldom migrates south, but simply descends into the lower, warmer valleys when it gets too cold in the highlands; the population may even shift back and forth on the slopes throughout the winter, staying just below the snow line.

Conversely, the **Red-breasted Nuthatch** normally remains on its territory year-round, so long as food is available. Although this little bird is not easily seen, because it usually stays high on the conifers, its call is one of the easiest to recognize and remember: a slow, nasal "nyak nyak nyak" call that sounds for all the world like a toy horn. Close up it has reddish underparts, a whitish throat, and a black crown that is divided by bold white eyelines. As might be expected from its name, it feeds primarily on nuts from conifers, which it extracts by hammering open the seeds after prying them out of the cones with its strong bill. It too stores food, wedging its supplies into bark crevices.

Pacific-slope Flycatchers seem to prefer the cooler canyon and shaded north slope; they are best identified by their very obvious call: "pee-ist" or "ps-seet ptsick." Chestnut-backed Chickadees are usually detected first by their drawling "tsic tsic tsic tyee check check" songs; they sport chestnut backs, black heads and bibs, and bold white cheeks. And the Winter Wren's song is never-ending, like tinkling trills and tumbling warbles; it is a tiny reddish bird with a short tail that usually stays hidden among the mossy roots and branches.

Other birds that are often found along or over the Cliff Nature Trail include the Red-tailed Hawk; Common Raven; Golden-crowned Kinglet; American Robin; Townsend's Solitaire; Brown Creeper; Nashville, Hermit, Yellow-rumped, and Wilson's Warblers; and Purple Finch. And by midsummer, when birds that have nested below or to the north of the monument are either wandering or en route south for the winter, several additional species can be expected: Turkey Vulture,

Band-tailed Pigeon, Mourning Dove, Rufous Hummingbird, Clark's Nutcracker, Varied Thrush, Pine Siskin, and Evening Grosbeak.

The Big Tree Trail (a 3-mile round trip) provides the visitor with a more complete perspective of the monument's biological diversity. Hiking the trail clockwise, one first experiences a relatively dry south slope with a Douglas-fir/oak community. Bushtits, Hutton's Vireos, and Nashville Warblers occur here during the breeding season. Not far beyond this area the trail enters an open white fir/Douglas-fir community where most of the same birds found along the Cliff Nature Trail can be expected. However, this area seems perfectly suited for the **Hermit Warbler,** a bird that usually stays in the upper canopy and requires patience for a good look. Locate it first by its rather distinct but somewhat varied song; according to my hearing it sings "che che che chezeee de-de" or "we we we wezeeee che-che," and sometimes it even adds an additional "che-che." In each case, the first three slightly ascending notes are followed by a buzzing note and two descending notes. Hoffmann described it somewhat differently: "The song varies greatly in different individuals; two common forms may be written 'wees-a wees-a wees-a wees' and 'tsip tsip tsip dee dee.'" Once located, it can be readily identified by its all-yellow face, black throat, whitish underparts, and gray back with two white wing bars.

The Hermit Warbler is one the area's better examples of a Neotropical migrant, nesting in the temperate zone and wintering in the tropics. Hermit Warblers arrive on their breeding grounds in late April, nest in May and June, stop singing by July, and depart by mid-August. Although this species' entire nesting range is limited to the dense conifer forests of Washington, Oregon, and Northern California, southbound migrants occur regularly as far east as Big Bend National Park, Texas, by the second week of August. They overwinter in the highland forests of northern Mexico and Central America. There they must find suitable habitat and food if they are to survive the winter and return to their breeding grounds the following spring.

By the time you reach the "Big Tree" and then follow the trail onto the north-facing slope, you will be entering a wetter environment that supports a few additional bird species. Agee and colleagues referred to this area as a "mesic white fir/Douglas-fir community." Although

Fig. 13. Hermit Warbler

most of the birds are duplicates of those already observed, there is a distinct increase in breeding Yellow-rumped Warblers, Townsend's Solitaires, Red-breasted Nuthatches, Brown Creepers, and juncos. Here too are nesting Hammond's Flycatchers and Gray Jays, and along the higher ridge, particularly along the Mt. Elijah Trail, are Olive-sided and Dusky Flycatchers, Mountain Chickadees, and Hermit Thrushes.

Yellow-rumped Warblers are the most numerous songbirds here, and their spirited songs, three "tsit" notes followed by an energetic trill on a lower pitch, ring through the forest. The bird is an active warbler, often flying from perch to perch, and so observing one requires less endurance than for its yellow-faced cousin. The Yellow-rumped male possesses a black chest and wings, white belly, and five yellow spots: on its cap, throat, sides, and rump. Like the Dark-eyed Junco, the yellow-throated Yellow-rumped Warbler of the western states, earlier known as "Audubon's," was lumped with the white-throated "Myrtle" Warbler of the East. The two forms interbreed where they overlap in the North; they are therefore considered the same species, and birders often refer to them as "butter-butts."

Gray Jays, perhaps more than any of the other breeding birds, are representatives of the northern forests. Their range extends throughout the boreal forests of Canada and Alaska. The Gray Jay is very different from the Steller's Jay, in both appearance and personality. It lacks a crest and adults are all gray with darker backs and whitish underparts; they often look like soft, fluffy toys. Grays often glide through the trees or down to inspect a hiker with muffled, low-pitched whistles and "cla cla cla" sounds. Like their crested cousins, however, they can be aggressive. They also store food for the winter, caching seeds, fruit, bugs, carrion, and even unattended nestlings on conifers. Their partially digested food is coated with sticky fluids from their mouths that help it to stick in place. This storage system permits them to remain on their territories throughout the cold winter months when little fresh food remains. It also allows Gray Jays to nest in late winter, when snow is still deep, before the arrival and/or growth of a new food supply.

Winter bird populations are only a fraction of what they are in spring and summer. Only a few of the more hardy species remain: Hairy Woodpecker, Red-breasted Sapsucker, Gray and Steller's Jays, Clark's Nutcracker, Mountain and Chestnut-backed Chickadees, Red-breasted Nuthatch, Brown Creeper, Golden-crowned Kinglet, Dark-eyed Junco, Purple Finch, and Evening Grosbeak.

In summary, the park's bird checklist includes eighty-four species, of which approximately fifty-seven are nesters. Of those fifty-seven species, none are waterbirds, eight are hawks and owls, and six are warblers.

BIRDS OF SPECIAL INTEREST

Pileated Woodpecker. This is the huge, red-crested "Woody Woodpecker" look-alike that has a bat-like flight and a loud "wuk wuk wuk" call.

Hairy Woodpecker. The most common woodpecker, it sports a white back and black wings and tail; males also have a red crown patch.

Steller's Jay. The monument's most obvious bird, it is usually present in the parking area, along the entrance trail, and at the cave entrance; it is all blue with a tall blackish crest and nape.

Gray Jay. Watch for this all-gray, noncrested jay in the old-growth forest; it too can be aggressive and curious.

Red-breasted Nuthatch. Usually detected first by its nasal "nyak" notes high in the conifers, it sports reddish underparts and a black-and-white head.

Yellow-rumped Warbler. It is best identified by its spirited songs, black chest, and five bright yellow spots: on its crown, throat, sides, and rump.

Hermit Warbler. A summer resident of the white fir/Douglas-fir forests, it has a yellow face, black throat, and whitish underparts.

Dark-eyed Junco. Watch for this little bird on the ground and in the undergrowth; it has a black hood, buff back and sides, and flashy white edges on its otherwise black tail.

Lava Beds National Monument, California

Rock Wrens are among the most numerous summer-resident birds at Lava Beds. Western Meadowlarks are the most vocal and Western Scrub-Jays the most obvious, but none surpass Rock Wrens in abundance. This little all-grayish-brown songster occurs at almost every rocky outcropping in the park, from the semiarid lowlands in the north to the tops of the highest cinder cones in the south. There is nowhere in the park where the rollicking songs of Rock Wrens cannot be heard throughout the summer months.

Rock Wren songs are extremely varied, ranging from a simple "cheer cheer cheer" to a series of phrases strung together in different combinations. John Terres, in *The Audubon Society Encyclopedia of North American Birds*, describes its song as "keree keree keree, chair chair chair, deedle deedle deedle, tur tur tur, keree keree trrrrr." And one June morning at Lava Beds, I recorded a Rock Wren that sang "teedly teedly teedly teedly, cheer cheer cheer, kreee kreee kreee."

Rather stocky birds with whitish eyelines, a hint of buff on their backs, and white-tipped tails, Rock Wrens have a habit of bobbing up and down with quick jerking motions. And when flying from one rocky perch to another, they will often spread their tails, bob up and down, and call sharp "tick-ear" notes from each new place. They will then commence to probe into every crack and crevice encountered, searching for insects, spiders, and other invertebrates with their long bills. Arthur Cleveland Bent, in *Life Histories of North American Nuthatches*,

Wrens, Thrashers, and Their Allies, reports that Oklahoma Rock Wrens prefer earthworms and grubs, and that of seventy-four Rock Wren stomachs examined in Utah, "30 contained 59 adult grasshoppers."

The Rock Wren's most fascinating habit, however, is its use of various materials to decorate the entrance to its nest. According to Bent, "one passageway of a hole in earth to nest was lined with 1,665 items, of which 492 were small granite stones, 769 bones of rabbits, fishes, birds, and nesting materials." The nest itself is lined with grasses, leaves, and other fine materials. Young of the year usually are out and about by early summer, and the majority of the park's Rock Wrens migrate to lower, warmer regions to the south for the winter months. However, just as soon as the first spring flowers appear, they are right back on their nesting grounds.

THE PARK ENVIRONMENT

Lava Beds, true to its name, is dominated by extensive lava flows and scattered cinder cones, lava tubes, spatter cones, and chimneys. Nearly three hundred lava tube caves, an unusual phenomenon, occur in the monument. They are all part of the Medicine Lake Volcano, a huge shield volcano that covers more than 750 square miles. The most recent lava flow in the monument is approximately 1,100 years old.

The entire monument slopes upward north to south from approximately 3,900 feet at the northeastern corner to 5,493 feet at the summit of Hippo Butte. Three rather distinct plant communities occur: big sagebrush-grasslands in the north; brushlands, dominated by mountain mahogany, bitterbrush, and waxy current; and pine forests, predominately ponderosa pine, with scattered areas of western juniper and white fir, and an undergrowth dominated by mountain mahogany and, in places, manzanita, on the higher slopes. Approximately 61 percent of the 46,500-acre monument is designated as wilderness.

The National Park Service operates a visitor center and campground at Indian Well, in the south-central portion of the park. The visitor center offers information services, exhibits, and a sales outlet: bird field guides and a checklist are available. Park Service interpretive activities include guided walks, evening campfire programs, and cave trips during the summer months. Schedules are posted in the campground and available at the visitor center. In addition, there is a self-guided nature trail, the historic half-mile-long Captain Jacks Stronghold Trail.

Additional information can be obtained from the superintendent, Lava Beds National Monument, P.O. Box 867, Tulelake, CA 96134; (916) 667-2282.

BIRD LIFE

Western Meadowlarks are most numerous in the sagebrush-grassland community, but their flutelike songs are commonplace in the higher brushlands as well. These were the first birds I encountered on entering the monument from the north. Their songs, a throaty gurgling sound, "tweet tweet, te-le-da, le-de-le," rang across the flats from every direction. A bird perched on a nearby shrub showed bright yellow underparts with a black V-shaped band across its chest. When I approached, it uttered low-pitched "tschuk" notes and flicked its stubby tail, showing white undertail feathers. It flew with rapid wingbeats, then glided to a more secure post some distance away. Through binoculars, I watched it stretch, flick its tail several times, and then proceed to sing its lovely song.

From Hospital Rock, I walked a couple miles on the Lyons Trail, which weaves through the sagebrush and patches of grasslands. Singing meadowlarks were common throughout, and several other birds were encountered: California Quail, Mourning Dove, Common Nighthawk, Horned Lark, Sage Thrasher, Spotted Towhee, Brewer's Sparrow, Brewer's Blackbird, Brown-headed Cowbird, and House Finch.

California Quail were obvious, either running down the trail or calling from open perches: emphatic "chi-ca-go," or, as Ralph Hoffmann describes their calls in *Birds of the Pacific States*, "come right here" and "where are you?" The park's only lowland quail, these are plump birds with prominent teardrop-shaped head plumes. Males are most colorful, with their coal black throats, edged with white, chestnut caps and sides, pale foreheads, and scaly bellies.

Sage Thrashers were surprisingly abundant, chasing one another here and there or sitting atop sagebrush shrubs for long periods of time. A few allowed a reasonably close approach. Their overall grayish

Fig. 14. Sage Thrasher

plumage included considerable buff on the rumps and tails, along with streaked underparts, whitish eyebrows and narrow wing bars, and tail feathers edged and tipped with white; their bicolored tails were especially evident in flight. Their songs blended with those of meadowlarks at first; they have some of the same fluty character. But Sage Thrashers sing a softer melody, not unlike that of the mocking-bird, a long series of rich warbling phrases, "with very little range in pitch and with constant repetition of one accented note," according to Hoffmann.

Male Sage Thrashers perform courtship flights over their territories. Fred Ryser Jr. describes their behavior in Birds of the Great Basin: A Natural History: "They fly in an undulating fashion, low over the sagebrush, in zigzag or circular paths, singing as they go. At the termination of a flight, they land with their wings upraised and flutter them briefly. As they continue to sing from a song perch, they repeatedly raise and lower their wings." Their wing-raising behavior is similar to the "wing-flashing" of their mockingbird cousins, although thrashers raise their wings smoothly, not in hitches.

Suddenly I was attracted to sweet but unmusical warbling sounds high overhead. It took me several minutes to find the source, a **Horned Lark** circling so high that I was forced to use binoculars to see it well. It was flying with its tail spread, showing the black-and-white pattern. Then it suddenly dropped like a rock, landing on the ground 150 feet or so in front of me. It took me several more minutes to maneuver into a position where I could see it well. It was a brightly marked male with a boldly marked head pattern: yellow throat and forehead, broad black facial stripes, and a black cap that extended into black tufts or "horns." Otherwise, except for a black throat and tail, with white outer tail feathers (evident only in flight), its back was all grayish brown, which blended in very well with the terrain. See the chapter on Olympic for further details about this bird's territorial display.

Throughout my walk, **Common Nighthawks** also cavorted overhead, chasing after insects. Occasionally one would dive steeply and produce a loud booming or "peent" sound, from its vibrating primaries, close to the ground and its nesting site. In normal flight, they often called nasal "spee-ick" notes. Nighthawks are members of the goatsucker family (Caprimulgidae), or nightjars. They are fairly large birds with long, pointed wings with a bold white slash. Although most abundant at dawn and dusk, during the breeding season and in migration they may fly during the daylight hours as well.

I also walked Captain Jacks Stronghold Trail, where I encountered most of the same birds found on the Lyons Trail. There were a few additions, however, principally due to the juniper trees at the picnic grounds near the parking area. I found four species nesting among the junipers: American Robin, Brewer's Blackbird, Bullock's Oriole, and

House Finch. A pair of Western Wood-Pewees was also present and singing; they probably were also nesting there. The American Robin nest was located against a trunk, about 15 feet above ground, and built primarily of grasses. Two oriole nests, woven from grasses, hung from the outer branches. The House Finch nest, also mostly of grass, had been built in a crevice made by overlapping branches. But most surprising was the Brewer's Blackbird nest, which had been built on a large branch about eight feet high. This all-black bird (males possess white eyes) usually nests on the ground or among sagebrush.

There were at least eight **Bullock's Orioles** near the parking area, chasing one another about the junipers or along the adjacent hillside. Three males sang their guttural, flutelike notes from the junipers or nearby elderberry shrubs. Truly gorgeous birds, males sport deep orange underparts and cheeks, coal black throats, caps, and backs, and large white wing patches. This species was lumped with the eastern "Baltimore" Oriole for several years, and called "Northern Oriole." It has since been split once again.

A few other birds were recorded on my walk: A Say's Phoebe was singing plaintive "pee-ee" songs from a perch near one of the many lava caves, where it probably was nesting. Pairs of Mourning Doves and Common Ravens passed overhead. Rock Wrens were plentiful. Western Meadowlark songs were all about. Spotted Towhees called drawling "chee-ee" notes from various thickets. A pair of Cassin's Finches flew away as I approached. And several Ring-billed Gulls and Red-winged Blackbirds, from nearby Tule Lake, passed overhead.

Farther along the roadway, I stopped at Canbys Cross parking area to see if the numerous junipers there contained as many nesting birds as I had found at Captain Jacks Stronghold. But except for a pair of Oak (plain) Titmice calling scratchy "ti-wee, ti-wee, ti-wee" notes, no other songbirds were detected. However, three **Barn Owls** flew out of the junipers next to the roadway when I approached on foot. Each flew only to other nearby junipers where they settled into hiding again. At least one of the three was a young bird, and so I assumed that I had found a family that was roosting near good hunting grounds (agricultural fields along Tule Lake are less than a mile away) rather than returning to nest sites on the lava cliffs. The park's resource manage-

Fig. 15. Bullock's Oriole

ment specialist Chuck Barat later told me that Barn Owls are common at Lava Beds; he found twenty nesting birds on the east side of the Petroglyph Cliffs on one occasion.

On my visit to Petroglyph Cliffs, Cliff Swallows were most numerous, and in the process of nest-building. I also found pairs of Red-tailed Hawks and American Kestrels nearby, and park interpreter Michele Moore told me that Prairie Falcons nest on these cliffs as well. She also pointed out that Barn, Great Horned, and Short-eared Owls can often be seen here at twilight, and that Bald and Golden Eagles and Rough-legged Hawks roost here in winter.

Another morning I walked the Cave Loop Road above the visitor center. It passes through typical brushlands and provides a good perspective on the birds that inhabit that community. The most obvious bird that morning was the **Western Scrub-Jay,** a long-tailed, non-crested blue bird with blackish cheeks, whitish eyelines and throat, and grayish-brown back. It also has loud, harsh "tschek, tschek, tschek" and "ker-wheek" calls. Scrub-jays stayed far enough away, for their presumed safety, along the Cave Loop Road, but at the nearby campground they were not so shy, coming into the campsite in search of handouts.

Several other birds were found on my morning walk: California Quail, Say's Phoebe, Violet-green and Barn Swallows, Purple Martin, Common Raven, Bewick's and Rock Wrens, Western Tanager, Western Meadowlark, Brewer's Blackbird, Brown-headed Cowbird, Spotted Towhee, Lazuli Bunting, Black-throated Sparrow, and House and Cassin's Finches.

Spotted Towhees were reasonably common, singing loud trills from various high posts. The male's coal black hood, with blood-red eyes, spotted back and wings, whitish underparts, and dark tail with large white spots near the tip, were most obvious. This large sparrow spends considerable time on the ground, searching for seeds and invertebrates. It has a habit of jumping backwards, dragging both feet, to clear away leaves in its search for food. The scratching sounds in the leaf litter can often be used in locating one of these brightly marked songbirds.

I added a few additional species in and about the campground: Ash-throated Flycatcher, Mountain chickadee, Oak Titmouse, Bushtit, and both Western and Mountain Bluebirds. Finding the two species of bluebirds was surprising because these lovely birds usually occur in slightly different habitats. Western Bluebird males are dark blue with a russet breast and flanks; they normally prefer mountain forests. Mountain Bluebird males are sky blue with lighter underparts; they normally prefer open sage flats with scattered trees. Both use natural cavities and deserted woodpecker nests for nesting.

During the fall, "flocks of Pinyon Jays can often be found around the visitor center and campground, and Townsend's Solitaire and American Robin numbers also seem to skyrocket in the juniper areas," according to Moore.

On another day I birded the pine forest on Mammoth Crater and walked into nearby Hidden Valley. Dark-eyed Juncos, Steller's Jays, American Robins, Rock Wrens, Townsend's Solitaires, and Hairy Woodpeckers were most numerous, more or less in that order. I also recorded a few Northern (Red-shafted) Flickers; an Olive-sided Flycatcher called loud "three-beers" from a high perch; a couple of Clark's Nutcrackers flew overhead; Mountain Chickadees called hoarse "chick-a-dee" notes from the high foliage; White-breasted and Pygmy Nuthatches were found on ponderosa pines in Hidden Valley; House Wrens called rambling songs from dense thickets; a pair of Western Bluebirds flew overhead; a male Yellow-rumped Warbler was singing from a tall pine; a singing Green-tailed Towhee on the far slope; a Cassin's Finch added its warbling song to the chorus; and a small flock of Pine Siskins flew overhead. Soaring birds included a lone Turkey Vulture and Cooper's Hawk and a pair of Red-tailed Hawks.

The most common of all these birds were the little **Dark-eyed Juncos**. A few sang from the treetops, their musical trills occasionally changing pitch in mid-song. Juncos found at ground level were feeding by jumping forward and pecking at various seeds and tiny invertebrates found on the ground. One brightly marked male sat at the top of a mountain mahogany for a long time, providing me with a wonderful view of its coal black hood, brownish back and sides, white underparts, and black tail with white outer tail feathers. A truly lovely bird!

It is impossible to discuss the birds of Lava Beds National Monument without also including a few waterbirds that are ever present along the park's northern boundary, at Tule Lake National Wildlife Refuge. During a one-hour stop at the East Wildlife Turnout, I recorded more than two dozen species. The majority of those were waterbirds: Eared and Western Grebes, American White Pelican, Double-crested Cormorant, Great Egret, White-faced Ibis, Canada Goose, Mallard, Gadwall, Cinnamon Teal, Lesser Scaup, Ruddy Duck, Sora, American Coot, Killdeer, Ring-billed Gull, and Forster's, Caspian, and Black Terns. Additional birds included Cliff and Barn Swallows, Marsh Wren, Song Sparrow, and Red-winged Blackbird.

Of all these, the **American White Pelican** was the most obvious. Its large size, all-white plumage, except for black wingtips evident in flight, huge orange to yellow bill, and habit of soaring in formations cannot help attracting one's attention. The Tule Lake pelicans are only visitors from nesting colonies at nearby Upper Klamath, Lower Klamath, and Clear Lake refuges. All the White Pelicans migrate south for the winter months, returning in late spring. An estimated 45,000 ducks, 2,600 Canada Geese, and thousands of marsh birds and shorebirds are raised at Tule Lake annually.

August to early November are the best months to see the millions of birds that pass through the refuge during their southbound migration. Peak numbers of nearly one million ducks and geese are usually present in early November, according to the refuge brochure. And the Klamath Basin refuges "host the largest wintering concentration of bald eagles in the lower 48 states. More than 500 of these majestic raptors are attracted to the refuges by the thousands of waterfowl that winter here, providing an abundant food supply."

Christmas Bird Counts, undertaken annually at Tule Lake, California, provide the best perspective on winter species. In December 1997, counters tallied 40,555 individuals of 84 species. The dozen most numerous of the 84 species, in descending order of abundance, included White-fronted Goose, Northern Shoveler, Canada Goose, American Wigeon, Red-winged Blackbird, Mallard, Northern Pintail, Tundra Swan, Canvasback, Snow Goose, Song Sparrow, and House Finch.

In summary, the park/refuge checklist of birds includes 225 species. Of those, sixty-nine are waterbirds (loon, grebes, pelican, cormorant, waders, waterfowl, crane, rails, and shorebirds), twenty-four are hawks and owls, and ten are warblers.

BIRDS OF SPECIAL INTEREST

American White Pelican. This huge, all-white bird, with its great orange to yellow bill and black wingtips (in flight), is common at Tule Lake.

California Quail. Watch for this plump bird running across or along the roads; it has a teardrop-shaped plume and males sport a black-and-white face pattern.

Barn Owl. Search the junipers along the entrance road for these "monkey-faced" owls; they nest in crevices on the lava cliffs.

Common Nighthawk. This is the long-winged bird flying over the sagebrush flats; notice the bold white slash on each wing.

Horned Lark. Common on the sagebrush flats, it is easily identified by its yellow, black, and white head pattern; flying birds show black tails with white outer tail feathers.

Western Scrub-Jay. This is the long-tailed, noncrested blue jay of the brushlands and juniper woodlands; it is common about the campground.

Rock Wren. Present at rocky outcroppings in summer, it is all grayish brown, bobs up and down, and has a "tick-ear" call.

Sage Thrasher. Watch for this gray-brown bird, with a streaked breast, among the sagebrush; its song is a mellow warble.

Western Meadowlark. This is the chunky yellow-breasted bird with a bold, V-shaped band across the chest; its song is a throaty, gurgling sound: "tweet tweet, te-de-da, le-de-le."

Spotted Towhee. Males sport coal black hoods with blood-red eyes, spotted backs and wings, and black tails with bold white spots.

Dark-eyed Junco. This is the common little black-headed bird with a brown back and sides and black tail with obvious white outer tail feathers.

Bullock's Oriole. Watch for this orange, black, and white bird about the junipers at Captain Jacks Stronghold parking area.

Lassen Volcanic National Park, California

Manzanita Lake was like a blue jewel in a forest green setting that late May morning. I followed the circle trail from the campground, alert for whatever birds might appear. The most abundant and most obvious was the perky Steller's Jay. It had been the first bird species seen on entering the park, and since then there had hardly been a time when one or several were not either visible or heard. Their loud, raspy "shaack, shaack, shaack" calls and a dozen other jay sounds seemed to permeate the environment. The bird's dark blue, almost metallic blue, body and blackish head, nape, and crest could hardly be mistaken for anything else. The Steller's Jay, more than any of Lassen's other birds, can properly be considered its best representative.

Steller's Jays are members of the crow (Corvidae) family, a group of extremely aggressive and highly intelligent birds. Opportunists, they eat a wide variety of foods. John Terres, in *The Audubon Society Encyclopedia of North American Birds*, reports that this jay forages "in treetops and on the ground, eats many acorns, pine seeds, some wild and cultivated fruit, beetles, wasps, bees, grasshoppers, caterpillars and moths, spiders, sow bugs, frogs, eggs and young of small birds; sometimes attacks and kills snakes." The Steller's Jay also caches foods and is able to transport small quantities of food in its esophagus.

Steller's Jays were nest-building during my late May visit, gathering huge billfuls of pine needles and grasses along the lakeshore. These were transported to dense stands of pines and cemented together with mud to form large, bulky nests, usually on a limb against the tree trunk. I watched a pair of courting birds behind the Loomis Museum: the male was bobbing up and down before his mate; then he jumped high in the air, landing on the opposite side, where he commenced his bobbing motion once again. He then began to circle his mate, bobbing all the while. On one occasion, he picked something off the ground and offered it to his mate; she accepted that token of his affection, perhaps sealing their mutual bond.

THE PARK ENVIRONMENT

Lassen Peak lies at the southern end of the Cascade Range and is one of only a few active volcanoes in the greater Pacific "Ring of Fire," a ring of volcanoes that encircles the Pacific Ocean. Lassen last erupted in 1915 when, according to the park brochure, "it blew an enormous mushroom cloud some 7 miles into the stratosphere. The reawakening of this volcano, which began as a vent on the flank of a larger extinct volcano known as Tehama, profoundly altered the surrounding landscape." Lassen Volcanic National Park was established in 1916.

The topography of the 106,372-acre park generally slopes upward, east to west, from 5,300 to 10,457 feet in elevation. The lower portion of the park is well forested and dotted with numerous lakes. It then rises to a plateau that covers the central portion and rises abruptly again to the steep, rugged crags of Lassen Peak.

Five distinct plant communities have been described for the park. The alpine community, found only above the tree line, contains low-growing pussy toes, pussy paws, and a number of hardy herbs. Next is the subalpine forest with whitebark pine and mountain hemlock, often in prostrate positions, which occur in scattered localities. The upper montane forest, a dense red fir forest with few understory plants, lies between 6,500 and 8,000 feet. And below 6,500 feet is a mixed conifer forest. This is the park's richest community, dominated by Jeffrey and sugar pines, incense-cedar, white fir, and Douglas-fir. Woody understory plants include gooseberries, currants, squawcarpet, ceanothus, and snowberry. In the eastern part of the park is a Jeffrey pine/western juniper woodland.

Fig. 16. Steller's Jay

Three additional communities are scattered throughout the forested zones: riparian areas of willows, thin-leaf alder, black cottonwood, and aspen occur along streamsides; meadows, with a variety of grasses and herbs; and mountain chaparral, which is dominated by greenleaf manzanita, huckleberry oak, chinquapin, and two species of ceanothus (tobacco brush and snow brush). The majority of the park is wilderness, accessible only by trails. A single road circles Lassen Peak between the northwestern and southwestern corners. The National Park Service operates the Loomis Visitor Center/Museum near the northwestern entrance at Mazanita Lake in summer. There can be found an information desk, exhibits, an orientation program, and a sales outlet: bird field guides and a checklist are available.

The visitor center also is the hub of the park's interpretive activities, which vary from nature walks and talks to demonstrations and special programs. The *Lassen Park Guide*, a newspaper available at the entrance stations and visitor center, contains a schedule of all interpretive activities.

Additional information can be obtained from the superintendent, Lassen Volcanic National Park, P.O. Box 100, Mineral, CA 96063-0100; (530) 595-4444.

BIRD LIFE

The grassy shoreline of Manzanita Lake provides nesting sites for a number of waterbirds. The largest and most obvious of these is the Canada Goose, distinctly marked with a black head and neck and snow white throat. I found eight adults, one pair with three goslings, resting and grazing on the grassy bank. They all moved into the lake as I approached; the youngsters were escorted into the safety of the dense rushes growing just offshore.

Mallards were most numerous, some in pairs and others in small bachelor flocks. Pied-billed Grebes were also plentiful along the lakeshore, often sinking straight down into the water to escape detection. Their strange, low hooting calls, "cow-cow-cow," echoed across the lake. I searched the patches of rushes for their floating nests, which grebes construct of grasses and mud, but to no avail. American Coots, all-black birds with large white bills, were also common along the lakeshore.

Several Buffleheads were also present on the lake. These small ducks, with snow white underparts and a large head patch, are among our most attractive waterfowl. The Manzanita Lake birds, however, were only visitors, according to park naturalist Steve Zachary. Although they once nested at Manzanita Lake, they are "now restricted to the park's backcountry lakes that don't receive the same high level of visitor use."

A trio of **Spotted Sandpipers** was present, as well, another species that disappeared from Manzanita Lake but may be nesting there again; a section of the lake was closed to boaters and fishermen. I watched these birds for a long time as they chased one another about the lake, landing on partially submerged logs to teeter back and forth in their characteristic manner. Then they would dash off again, flying stiff-winged in a sporadic flight-and-glide manner to the next landing spot.

The Spotted Sandpiper, one of nature's most fascinating experiments, is one of the very few bird species in which the female is dominant. She will entice two or more males into her harem to incubate the eggs and care for the youngsters. She arrives on the breeding grounds first and selects and defends her territory and her mates. This example of avian polyandry is an apparent success story because Spotted Sandpipers are among our most widespread and abundant shorebirds.

Common Mergansers are only occasional visitors to Manzanita Lake, but they breed at many of the park's backcountry lakes. They use tree cavities as well as holes in the bank or under roots and clumps of grass. Terres reports that a New Hampshire bird attempted to nest in a house chimney, and members of this species have even nested in Arizona's Montezuma Castle cliff dwelling. Slightly larger than Mallards, male mergansers also possess bright green heads. But the resemblance stops there. Mergansers are predators that dive completely underwater and swim down their prey. They possess hooked bills with serrated edges so they can catch and hold onto the slippery fish. Their bills usually are bright red, and both sexes possess white breasts.

Drakes are most distinguished with their black-and-white bodies; hens possess chestnut heads and grayish bodies.

An extensive area of mountain chaparral occurs along the southwestern corner of Manzanita Lake, providing rather choice habitat for four very different bird species: Mountain Quail, Willow Flycatcher, Yellow Warbler, and Fox Sparrow. Although **Mountain Quail** also frequent the park's red fir forest, they spend considerable time at chaparral patches. One of the few quail species that undergoes an altitudinal migration, the Mountain Quail apparently uses these areas in spring because they are often one of the first areas to be snow free. It also feeds on ripe manzanita berries in summer.

Mountain Quail are extremely shy and normally run away instead of taking flight when disturbed. Therefore, it is difficult to get a good look at these large quail. Once found, however, they are easy to identify by their long, straight black plumes, gray chests, and chestnut throats and flanks, with bold white bars. Often they are detected first by their loud calls, which Ralph Hoffmann, in Birds of the Pacific States, refers to as a "rapidly repeated 'kup kup kup,'" and in late summer when a covey of Mountain Quail are gathered under cover, there issues a medley of clucking, mewing, and whining sounds, mixed with harsh squawks and a 'ka-yak, ka-yak' like a guinea fowl's."

Two Willow Flycatchers were perched on scattered white firs, singing loud "fitz-bew" songs. Yellow Warblers also were singing from the conifers, a cheerful "tseet-tseet-tseet sitta-sitta-see." One brightly marked male came close enough so that I was able to see its all-yellow underparts streaked with chestnut. And I watched a particularly vocal Fox Sparrow, moving about its apparent territory, singing loud, extensive songs from various high points. This is a large, rust-colored sparrow with breast streaks that converge into a dark spot.

Other birds recorded on my Manzanita Lake Trail walk included the White-headed and Hairy Woodpeckers, Northern Flicker, Olive-sided and Hammond's Flycatchers, Western Wood-Pewee, Mountain Chickadee, Red-breasted and White-breasted Nuthatches, Brown Creeper, Ruby-crowned Kinglet, Hermit Thrush, American Robin, Warbling Vireo, Hermit Warbler, Western Tanager, Dark-eyed Junco, Song

Sparrow, Brewer's Blackbird, Brown-headed Cowbird, and Evening Grosbeak.

Zachary surveyed the birds around Manzanita Lake for several years, recording a total of eighty-nine species, of which thirty-two were found to nest. His list of breeders added seven species: Common Nighthawk, Red-breasted Sapsucker, Tree Swallow, House Wren, American Dipper, and Orange-crowned and Wilson's Warblers. Park naturalist Scott Isaacson told me that "common nighthawks can be abundant over the lake on summer evenings."

But of all the birds that frequent Manzanita Lake, none is as colorful as the male **Western Tanager.** This is a canary yellow bird with a black tail and wings, with yellow wing bars, and a bright red head; females are yellowish with darker wings. According to Robert Milne, in the out-of-print booklet *Birds of Lassen Volcanic National Park*, Western Tanagers are a "common summer resident of the pine-fir forests at the lower elevations and may usually be seen foraging for insects among lichen-covered branches or around the shrub-lined lake borders." They often sit high on tall trees and sing a song that sounds somewhat like that of a robin but hoarser and seldom containing more than four or five phrases. Hoffmann describes the song best: "It is made up of short phrases with rising and falling inflections 'pir-ri pir-ri pee-wi pir-ri pee-wi.'" They also have a distinct "prit-it" or "pri-ti-tick" call.

Dark-eyed Juncos were common on the ground and among the undergrowth, where they were searching for seeds. Territorial birds were singing from high posts, even from the very tops of conifers. Their musical trills rang out across the forest and lake. Males sport coal black hoods, brownish backs, wings, and tails, pinkish sides, and white underparts and flashy outer tail feathers; females are duller versions. This bird was once known as "Oregon Junco" but was lumped with all the other Dark-eyed Juncos.

Watch for the **American Dipper** along the park's swift streams, such as Manzanita and Hat Creeks, and at King's Creek Falls. Dippers, sometimes called "water ouzels," are our only truly aquatic songbirds. These plump, dark gray birds spend much of their time underwater, feeding on water insects and their larvae found on the stream bottoms.

They have adapted to these habitats to such an extent that they actually can swim through the swift currents, to as much as 20 feet below the surface, using their short, rounded wings as flippers and stubby tails as rudders. They possess special adaptations for this unique behavior: much larger oil glands than other songbirds, used to waterproof their feathers, and scales that close their nostrils when underwater. Their name is derived from the habit of bobbing up and down on bent knees.

At Hat Lake, the surrounding riparian vegetation contained a number of typical riparian birds: Calliope Hummingbird, Downy Woodpecker, Willow Flycatcher, Tree Swallow, Warbling Vireo, Wilson's and MacGillivray's Warblers, and White-crowned and Song Sparrows. Most numerous were the **Wilson's Warblers**, singing rapid songs, which Wayne Peterson, in *The Audubon Society Master Guide to Birding*, calls a "staccato chatter dropping in pitch at end: 'chi chi chi chi chet chet.'" Males were especially active, singing from various posts about their territories, and responding immediately to any trespasser or atypical sound. Two males came within a few feet of me when I spished; their all-yellow underparts and solid black caps were obvious.

SOUTHWESTERN AREAS

The park's southwestern entrance area, including the Lassen Chalet area and adjacent picnic grounds, supports many of the same birds that occur near Manzanita Lake. Betty and I watched a pair of **White-headed Woodpeckers** searching for insects on an old snag very near our picnic table. They seemed oblivious to our admiring stares, searching the crevices and holes for whatever invertebrate prey they could find. The White-headed is a small woodpecker with contrasting plumage: all black except for its snow white forehead, face, throat, and wing patches; males also sport a bright red patch on the back of the head.

Woodpeckers, especially the Hairy and White-headed and Northern Flicker, were present throughout the forest. I also found a Pileated Woodpecker, a huge bird with a bright red crest, along the Lily Pond Nature Trail; a Red-breasted Sapsucker along the stream below Lassen Chalet; a Williamson's Sapsucker near the parking area at Summit Lake; and a Black-backed Woodpecker on the far side of Summit Lake.

Fig. 17. Western Tanager

I walked down the steep slope below the picnic area to the stream, and then walked a loop through the adjacent meadow, returning to the picnic area via the chalet. A Blue Grouse was booming on the far slope; the deep, hollow "broo" sounds first increased in volume and then diminished. Several other birds were in full song: Olive-sided Fly-catcher, Western Wood-Pewee, Mountain Chickadee, Red-breasted Nuthatch, House Wren, American Robin, Yellow-rumped Warbler, Dark-eyed Junco, and Cassin's Finch. A pair of Red Crossbills flew over, calling loud "jip jip" notes in flight. And just below the chalet, a pair of Tree Swallows began diving at me; apparently they had a nest in a nearby tree.

One of the most numerous birds that day was the brightly marked **Yellow-rumped Warbler.** Males were especially colorful with their coal black breasts and sides, white underparts, and five bright yellow spots: on their caps, throats, sides, and rumps. Several sang spirited songs, which Hoffmann describes as "a rather characterless succession of notes, beginning 'tsit, tsit, tsit,' followed by a loose, less energetic trill on a lower pitch." I also watched both males and females dash out after passing insects from various perches, sometimes even diving to the ground to capture a particularly choice insect. This warbler was known as "Audubon's Warbler" until it was lumped with the eastern "Myrtle Warbler," after the two forms were found to interbreed where their ranges overlapped in the North.

I found several Steller's Jays and a pair of American Robins feeding on flying carpenter ants that were emerging from holes in an old snag. The birds perched on an adjacent limb and flew down to take a winged ant whenever one appeared. They did not bother any of the more abundant nonwinged ants; I wondered if the winged versions were tastier or more nutritious.

We also watched a Golden Eagle that was soaring high overhead. It was an adult bird with all-dark plumage, and, with binoculars, I could see the golden sheen of its head when it circled. Not long afterward, we found an immature Golden Eagle soaring above the highway; it still had a broad whitish tail band and underwing patches. Other large soaring birds seen during our visit included several Turkey Vultures, occasional Red-tailed Hawks, and a lone Osprey over Summit Lake.

Zachary told me that Ospreys, as well as Bald Eagles, are being found in the park more often now than a few years ago. Populations of both species seriously declined throughout the West while DDT was in use, but they have increased ever since that deadly chemical was banned in 1972. Today, both of these charismatic birds nest in the area and fish some of the park's backcountry lakes.

Eagles and Turkey Vultures are often confused because of their similar appearance, all-dark plumage, and large size. However, Turkey Vultures have bare, red heads, compared with the large, feathered heads of eagles; vultures possess bicolored wings, compared with the all-dark eagle wings, and they fly with their wings in a shallow V pattern and rock slightly from side to side; eagles do not rock in flight, nor do they hold their wings at such an angle.

The park highlands about Lassen Peak's rocky crags support only a few bird species. Tree-line species include the Clark's Nutcracker, Mountain Chickadee, Mountain Bluebird, Townsend's Solitaire, and

Cassin's Finch. The Gray-crowned Rosy-Finch, which once nested above the tree line, has apparently disappeared, according to Zachary. He has not found it in recent years. Prairie Falcons, however, can often be seen in the alpine area, and Rock Wrens nest among the open talus slopes.

The **Clark's Nutcracker** nests in this subalpine community in late winter and spends the summer months at somewhat lower elevations; it is present at the Lassen Chalet year-round. This black-and-white Corvid has a harsh "kra-a-a" or "chaar" call that can be heard for a considerable distance. Its scientific name is *Nucifraga columbiana*; the genus is Latin for "nut-breaker," after its habit of cracking and eating conifer seeds. The species name was derived from the Columbia River, the place where William Clark, of the Lewis and Clark Expedition, was the first to collect one in about 1804. See the chapter on Crater Lake for additional information about this bird's food caching behavior.

The majority of the park's nesting songbirds leave their breeding grounds by late summer. Many of these are Neotropical migrants that spend their winters in Mexico, Central America, or, occasionally, in South America. Fewer species either remain on site all winter or move to lower elevations only during extreme cold periods. Only the hardiest species, such as the Canada Goose, woodpeckers, Steller's Jay, Common Raven, Clark's Nutcracker, Mountain Chickadee, Red-breasted Nuthatch, and Brown Creeper can be expected on their breeding grounds year-round.

In summary, the park's checklist of birds includes 195 species, of which 89 are known to nest. Of those eighty-nine nesters, fifteen are waterbirds, nine are hawks and owls, and seven are warblers.

BIRDS OF SPECIAL INTEREST

Common Merganser. This fish-eating duck occurs on backcountry lakes; males possess green heads and white underparts, while females have reddish heads and white throats.

Mountain Quail. Watch for this large quail with a long, straight plume in areas of manzanita or along the roadsides in red fir forest areas.

Spotted Sandpiper. Fairly common at all the park lakes, breeding birds possess heavily spotted breasts and teeter back and forth.

White-headed Woodpecker. This is a small, all-black woodpecker with a white face and wing markings.

Steller's Jay. The park's most common bird, it is all royal blue with a blackish crest and back and loud, aggravating shrieks.

Clark's Nutcracker. Watch for this black-and-white bird in the higher elevations; it has loud, rasping "kra-a-a" or "chaar" calls.

American Dipper. This is the plump, dark gray bird that lives along swift streams; it actually swims underwater and feeds on the stream bottoms.

Yellow-rumped Warbler. A common forest bird with black, yellow, and white markings, including five yellow spots: on its cap, throat, sides, and rump.

Wilson's Warbler. Watch for this little yellow bird along streams; males possess solid-black caps.

Western Tanager. Males are gorgeous birds with canary yellow under-parts, backs, and wing bars, black wings, and bright red heads.

Dark-eyed Junco. This is the little ground-feeding bird with a black hood and tail, with flashing white outer tail feathers, brownish back, and pinkish sides.

References

Agee, James K., and Jane Kertis. 1986. *Vegetation Cover Types of the North Cascades.* Seattle: National Park Service Cooperative Park Studies Unit, University of Washington.

Agee, James K., Laura Potash, and Michael Granz. 1990. *Oregon Caves Forest and Fire History.* Seattle: National Park Service Cooperative Park Studies Unit, University of Washington, 90–91.

American Bird Conservancy. 1997. *All the Birds.* New York: HarperCollins Publishers.

American Ornithologists' Union (AOU). 1998. *Check-list of North American Birds,* 7th ed. Washington, D.C.: American Ornithologists' Union.

Arno, Stephen F. 1984. *Timberline Mountain and Arctic Forest Frontiers.* Seattle: The Mountaineers.

Bellrose, Frank C. 1976. *Ducks, Geese, and Swans of North America.* Harrisburg, Pa.: Stackpole Books.

Bent, Arthur Cleveland. 1963. *Life Histories of North American Flycatchers, Larks, Swallows, and Their Allies.* New York: Dover Publications.

———. 1963. *Life Histories of North American Wood Warblers.* New York: Dover Publications.

———. 1964. *Life Histories of North American Jays, Crows, and Titmice.* New York: Dover Publications.

———. 1964. *Life Histories of North American Thrushes, Kinglets, and their Allies.* New York: Dover Publications.

———. 1964. *Life Histories of North American Nuthatches, Wrens, Thrashers, and Their Allies.* New York: Dover Publications.

Brockman, C. Frank. 1979. *Trees of North America*. New York: Golden Press.

Brooks, Paul. 1980. *Speaking for Nature*. San Francisco: Sierra Club Books.

Burt, William Henry, and Richard Philip Grossenheider. 1952. *A Field Guide to the Mammals*. Boston: Houghton Mifflin.

Butcher, Devereux. 1956. *Exploring Our National Parks and Monuments*. Boston: Houghton Mifflin.

Carson, Rachel. 1962. *Silent Spring*. Boston: Houghton Mifflin.

Chadwick, Douglas H. 1990. "The Biodiversity Challenge." *Defenders* (May/June): 19–31.

Chapman, Frank M. 1966. *Handbook of Eastern North American Birds*. New York: Dover Publications.

Clark, William S., and Brian K. Wheeler. 1987. *A Field Guide to Hawks of North America*. Boston: Houghton Mifflin.

Council on Environmental Quality and Department of State. 1980. *The Global 2000 Report to the President*. Washington, D.C.: GPO.

Dawson, William Leon. 1909. *The Birds of Washington*. The Occidental Publishing Co.

de la Torre, Julio. 1990. *Owls: Their Life and Behavior*. New York: Crown Publishers.

Diamond, Antony W., Rudolf L. Scheiber, Walter Cronkite, and Roger Tory Peterson. 1989. *Save the Birds*. Boston: Houghton Mifflin.

Dunn, Jon L., and Kimball L. Garrett. 1997. *A Field Guide to Warblers of North America*. Boston: Houghton Mifflin.

Ehrlich, Paul R., David S. Dobkin, and Darryl Wheye. 1988. *The Birder's Handbook*. New York: Simon & Schuster.

Ehrlich, Paul R., David S. Dobkin, and Darryl Wheye. 1992. *Birds in Jeopardy*. California: Stanford University Press.

Farb, Peter. 1963. *Face of North America*. New York: Harper & Row.

Farner, Donald S. 1952. *The Birds of Crater Lake National Park*. Lawrence: University Press of Kansas.

Follett, Dick. 1979. *Birds of Crater Lake National Park*. Crater Lake Natural History Association. Crater Lake National Park, Ore.

Franklin, Jerry F., and C. T. Dryness. 1973. *Natural Vegetation of Oregon and Washington*. General Technical Report PNW-8. Portland, Ore.: USDA Forest Service.

Franklin, Jerry F., William H. Moir, Miles A. Hemstrom, Sarah E. Greene, and Bradley G. Smith. 1988. *The Forest Communities of Mount Rainier National Park*. Washington, D.C.: U.S. Department of the Interior, National Park Service.

Freeman, Judith. 1986. "The Parks as Genetic Islands." *National Parks* (January/February): 12–17.

Getty, Stephen R. 1993. "Call-notes of North American Wood Warblers." Birding (June): 159–68.

Graham, Frank, Jr. 1990. "2001: Birds That Won't Be with Us." *American Birds* (winter): 1074–81, 1194–99.

Gruson, Edward S. 1972. *Words for Birds: A Lexicon of North American Birds with Biographical Notes.* New York: Quadrangle Books.

Guggisberg, C. A. W. 1970. *Man and Wildlife.* New York: Arco Publishing.

Halle, Louis J. 1947. *Spring in Washington.* New York: Harper & Brothers, Publishers.

Harrison, Peter. 1985. *Seabirds: An Identification Guide.* Boston: Houghton Mifflin.

Headstrom, Richard. 1951. *Birds' Nests of the West: A Field Guide.* New York: Ives Washburn.

Hoffmann, Ralph. 1927. *Birds of the Pacific States.* Boston: Houghton Mifflin.

Hoose, Phillip M. 1981. *Building an Ark.* Covelo, Calif.: Island Press.

Hutto, Richard L. 1988. "Is Tropical Deforestation Responsible for the Reported Declines in Neotropical Migrant Populations?" *American Birds* (fall): 375–379.

Kaufman, Kenn. 1983. "Yellow-rumped Warbler." In *The Audubon Society Master Guide to Birding,* edited by John Farrand Jr. New York: Alfred A. Knopf.

Kitchin, E. A. 1939. "Birds of Mt. Rainier National Park." In *Mt. Rainier Park Nature Notes,* nos. 3 and 4.

Knox, Margaret L. 1990. "Beyond Park Boundaries." *Nature Conservancy* (July/August): 16–23.

Kuntz, Robert C., II, and Reed S. Glesne. 1993. *Stehekin Valley Vertebrate Inventory.* Seattle: National Park Service, Pacific Northwest Region.

Leopold, Aldo. 1966. *A Sand County Almanac.* New York: Oxford University Press.

Levy, Sharon. 1993. "A Closer Look: Marbled Murrelet." *Birding* (December): 421–425.

Line, Les. 1993. "Silence of the Songbirds." *National Geographic* (June): 68–91.

Matthiessen, Peter, and Ralph S. Palmer. 1967. *The Shorebirds of North America.* New York: Viking Press.

Milne, Robert C. 1966. *Birds of Lassen Volcanic National Park.* Mineral, Calif.: Loomis Museum Association.

Moorhead, Bruce B. 1994. *The Forest Elk.* Seattle: Northwest Interpretive Association.

National Audubon Society. 1983. "The Ninety-eighth Christmas Bird Count." *American Birds.* New York: National Audubon Society.

National Fish and Wildlife Foundation. 1990. "Proposal for a Neotropical Migratory Bird Conservation Program." National Fish and Wildlife Foundation. Photocopy. Washington, D.C.

National Geographic Society. 1999. *Field Guide to the Birds of North America.* 3d ed. Washington, D.C.: National Geographic Society.

Palmer, Ralph S. 1962. *Handbook of North American Birds.* Vol. 1. New Haven: Yale University Press.

————. *Handbook of North American Birds.* Vol. 4. New Haven: Yale University Press.

Peterson, Roger Tory. 1961. *A Field Guide to Western Birds.* Boston: Houghton Mifflin.

Peterson, Wayne R. 1983. "Winter Wren; Swainson's Thrush; Wilson's Warbler." In *The Audubon Society Master Guide to Birding,* edited by John Farrand Jr. New York: Alfred A. Knopf.

Pough, Richard H. 1957. *Audubon Western Bird Guide: Land, Water, and Game Birds.* Garden City, N.Y.: Doubleday & Co.

Rands, Michael, and Martin Kelsey. 1994. "Call to Action." *American Birds* (spring): 36–48.

Reader's Digest. 1985. *Our National Parks.* New York: The Reader's Digest Association.

Rich, Terry. 1989. "Forests, Fire, and the Future." *Birder's World* (June): 10–14.

Richmond, Jean. 1985. *Birding Northern California.* Walnut Creek, Calif.: Mt. Diablo Audubon Society.

Robbins, Chandler S., Bertell Bruun, and Herbert S. Zim. 1983. *A Guide to Field Identification: Birds of North America.* New York: Golden Press.

Robbins, Chandler S., John R. Sauer, Russell S. Greenberg, and Sam Droege. 1989. "Population Declines in North American Birds That Migrate to the Neotropics." *Population Biology* 86: 7658–62.

Ryser, Fred A., Jr. 1985. *Birds of the Great Basin: A Natural History.* Reno: University of Nevada Press.

Schirato, Greg. 1994. "Status Report on Harlequin Ducks in Washington, 1994." Photocopy. Olympic National Park, Wash.

Shafer, Craig L. 1990. *Nature Reserves Island Theory and Conservation Practice.* Washington, D.C.: Smithsonian Institution.

Sharpe, Fred A. 1993. "Olympic Peninsula Birds: The Songbirds." Unpublished manuscript. Olympic National Park, Wash.

Sherwonit, Bill. 1995. "Striking a Balance." *National Parks* (January/February): 27–31.

Smithson, Michael. 1947. *Olympic Ecosystems of the Peninsula.* Helena, Mont.: American & World Geographic Publishing.

Stewart, Charles. 1988. *Wildflowers of the Olympics and Cascades.* Port Angeles, Wash.: Nature Education Enterprises.

Stoltenburg, William. 1991. "The Fragment Connection." *Nature Conservancy* (July/August): 19–25.

Sutton, Ann, and Myron Sutton. 1974. *Wilderness Areas of North America.* New York: Funk & Wagnalls.

Terborgh, John. 1989. *Where Have All the Birds Gone?* Princeton: Princeton University Press.

———. 1992. "Why American Songbirds Are Vanishing." *Scientific American* (May): 98–104.

Terres, John K. 1987. *The Audubon Society Encyclopedia of North American Birds.* New York: Alfred A. Knopf.

Udall, James R. 1991. "Launching the Natural Ark." *Sierra* (September/October): 80–89.

van Vliet, Gus. 1993. "Flyways and the Marbled Murrelet." Unpublished report. Olympic National Park, Wash.

Wahl, Terry, and Dennis R. Paulson. 1973. *A Guide to Bird Finding in Washington.* Bellingham, Wash.: Whatcom Museum Press.

Wauer, Roland H., and Terrell Johnson. 1981. "La Mesa Fire Effects on Avifauna—Changes in Avian Populations and Biomass." Paper presented at La Mesa Fire Symposium, Los Alamos, N. Mex., October 6–7. Los Alamos, N. Mex.: Los Alamos National Laboratory.

Wilcove, David. 1990. "Empty Skies." *Nature Conservancy* (January/February): 413.

Wilhelm, Eugene J., Jr. 1961. *Common Birds of Olympic National Park.* Olympic Natural History Association. Olympic National Park, Wash.

Wilson, Edward O. 1992. *The Diversity of Life.* Cambridge, Mass.: Harvard University Press, Belknap Press.

Youth, Howard. 1992. "Birds Fast Disappearing." In *Vital Signs 1992.* Worldwatch Institute. New York: W. W. Norton.

Zwinger, Ann H., and Beatrice E. Willard. 1972. *Land above the Trees.* Tucson: University of Arizona Press.

Index